M000167358

T H

F U T

O

Y O

THE FUTURE OF YOU

CAN YOUR IDENTITY
SURVIVE 21st-CENTURY
TECHNOLOGY?

TRACEY FOLLOWS

Elliott&Thompson

First published 2021 by
Elliott and Thompson Limited
2 John Street
London WC1N 2ES
www.eandtbooks.com

ISBN: 978-1-78396-545-8

Typesetting by Marie Doherty
Printed in the UK by CPI Group (UK) Ltd,
Croydon, CR0 4YY

To Teddy & Emelda May

CONTENTS

INTRODUCTION: DISTRIBUTING YOU

I am not who I say I am. At least, not according to Facebook. A few years ago I opened an email from Facebook that began 'Dear Byron'. It was an email to me at my usual email address but the name of the person it was addressed to was Byron Loweth.

This continued for several weeks, with notifications from Facebook popping up to show me the latest status updates from my thirty-year-old male friends. Everyone seemed to be having a lovely time. The problem was I didn't recognise any of them. They weren't my friends, they were Byron's friends. I soon suspected my Facebook had been hacked. My login details didn't seem to work so I opted to lock my account, reset my email password, and then login afresh to Facebook. But the process required that I upload some personal identification, with a range of options offered such as a driving licence or a passport. I wondered if this might be a phishing scam but having satisfied myself that it all seemed above board, I scanned in a copy of my passport, and went off to make a cup of tea while I waited for Facebook to tell me it had unlocked my account.

I returned to find a message that said my account could not be unlocked. Having scrutinised my passport, Facebook had concluded this wasn't me. Wait, what? What did they mean this wasn't me? It was my Facebook account, with photos of me all over my feed, that matched the one on my passport. If not me, who did they think I was?

To this day, I have not been able to unlock my Facebook account and I can only assume that if I have any presence on the site at all, I do so in a semi-deactivated state of social limbo. But it got me thinking. I had been asked to prove my identity in order to access a website, and I had uploaded a copy of a document designed to give me access to whole countries – proof of identity that I had to carry around with me twenty-four hours a day when I lived abroad, and which you have to go to quite some trouble to replace should it be lost. But Facebook didn't recognise it, and it didn't recognise me. The automated system of a technology platform had decided that my government-issued paperwork didn't prove who I said I was – and there was no human within the system to whom I could appeal.

I began to worry. If a technology platform could set itself up as the authority when it comes to verifying my identity, what else might its algorithms – and those of all my other online service providers – be deciding about me? If I could be told that I wasn't me, I felt like question marks could be raised over every other aspect of my existence. And I started to wonder who was in charge.

'Who am I?' is one of the most fundamental questions that we can ask ourselves. It seems such a simple question to pose, but philosophers have wrestled with possible answers for centuries. And they are still arguing about it today. In 2013, David Bourget from the University of Western Ontario and David Chalmers from New York University conducted a survey of the views of just under two thousand contemporary professional philosophers and found that opinion was split on the nature of personal identity. Just over a third of respondents had a psychological view of personal identity – that I am who I am for as long as I remain capable of psychological processes such as thinking, believing, desiring, remembering and choosing. Just under 17 per cent had a biological view – that I am who I am for as long as my body continues to exist. And 12 per cent opted for the so-called 'further-fact view' – that who I am depends on more than my continued psychological and biological existence. And the rest backed various other unspecified theories.

If eminent philosophers can't agree on a universal theory of personal identity, then I am certainly not going to venture one. This book is not going to offer a philosophical analysis of identity; there are plenty of very good reads on that from philosophers past and present. Nor is it going to be about identity in the sense of 'identity politics' – the way in which people choose to identify themselves in all sorts of different ways that have become increasingly politicised such as nationality, race and gender. There is plenty to be said about movements such as #metoo and #blacklivesmatter – indeed, the media seem to cover them 24/7 – but they revolve around quite a narrow view of identity as something associated solely with the colour of our skin, our gender, class or sexuality. The kind of identity that I want to talk about is a far broader, richer concept.

I believe that the notion of identity has become more important than ever thanks to the disruptive effect of technology. After my experience with Facebook, it has dawned on me that a significant shift is taking place. In the past I always assumed that I had control over my own identity via my state-issued credentials – that they represented a mutually agreed proof of identity between myself and the state. And to a large extent that was true. Years ago, what you considered to be 'you' existed almost entirely in the analogue world. There were certain key documents and information linked to your identity held by the government and a few other official institutions such as your bank. But mostly what defined you as 'you' existed in the real world, under your control – your photographs, your receipts, your physical body, your work, your interactions with others. Over the last twenty years or so, however, with the rise of technology platforms like Facebook, we live much more of our lives online, relying on machines to verify, authenticate, validate and connect us under their terms of service rather than ours. It makes me wonder if we are in danger of losing control over who we are.

In this digital, data-driven, internet-connected world that we now find ourselves in, our identity has become distributed throughout a network of online sign-ins and subscriptions, through social platforms and work productivity apps, through online bank accounts and shopping carts that hold our credit card details, and through office entry systems that can verify who we are from the sound of our voice or a quick scan of our face. The interconnectivity of the digital world has great benefits; people can share ideas, collaborate on projects and learn from each other much more easily, and the more connected we are, the more innovative we are. But equally, the more connected we are, the more vulnerable we are. Now that our identities are distributed across the web, more people can gain access to our personal information, views, contacts and ideas, which has helped to drive an increase in identity fraud. In fact, synthetic identity fraud – the blending of real and fake information to invent a person who can access services and then disappear – is one of today's fastest growing forms of fraud.

Technology has caused us to think differently about who we are. The proliferation of social media platforms has given us greater opportunity to shape and share our personal identity. But they have also made it possible for us to present different personas via different media and distribute our 'self'. In January 2020 a hugely successful meme was born when the country singer Dolly Parton was invited to take part in what became known as 'the social media challenge', creating a grid of four photographs that would make perfect profile pictures for LinkedIn, Facebook, Instagram and Tinder accounts. Each of the images she uploaded for the challenge were quite distinct from each other: LinkedIn Dolly is pictured looking serious and professional in a work suit and an updo; Facebook Dolly is shown wearing a friendly girl-next-door Christmas sweater; Instagram Dolly is depicted in arty black and white, wearing a pair of jeans and carrying her guitar; and Tinder Dolly is dressed in a playboy bunny outfit – what else? And underneath the grid of images, she had written: 'Get you a woman who can do it all!'

As the meme took off, millions of people – celebrities and non-celebrities alike – proceeded to post their own versions, much to the amusement of their families, friends and followers. But behind the fun there was a serious point about the way in which most of us choose to showcase different personas for different contexts. We are expected to present different versions of ourselves not only to attract attention but to fit in. It's almost as if identity is increasingly about *performing* a role – and not just one role but many. In some ways this means that we are gaining a huge amount of control in terms of how we portray ourselves, with exciting new avenues appearing all the time for the exploration and creation of our identity. It also offers a challenge to our traditional notions of a single coherent identity, inviting us to embrace multiple personas and more fluid concepts of identity. But I also fear that the same technology could see us lose control of our identities entirely.

Take the social media services to which we enthusiastically upload so many of our photographs. Most of us probably have no idea what their terms of service are. While you technically still own the photos, you are also agreeing that the platform has the right to use them for as long as they are stored there. And if you aren't happy with that, it may not be enough just to delete your content if others have chosen to share and repost it. In theory these social media platforms could make use of your images in any way they like – whether you feel comfortable about it or not.

We are unthinkingly handing over huge amounts of our personal identity all the time. With a simple tick of a box – whether we're agreeing to terms and conditions that we haven't read or opting for some default privacy settings – we give permission to online platforms to build up an increasingly detailed picture of who we are: what we browse, what entertainment we like, who our friends are, where we live, when we get up in the morning and where we are at any minute of any given day.

In a 2018 discussion paper for Harvard Law School, researcher Dan Svirsky revealed the results of an experiment that he conducted to understand our attitudes towards online privacy. In the experiment, he first asked respondents to answer an online survey in exchange for a payment of $0.02, but offered them an extra $0.50 if they would also agree to share their Facebook profile. When faced with such a direct trade-off, the majority of participants chose not to disclose details about their Facebook profile. And the same was true for participants offered an extra payment of as much as $5, which would seem to suggest that we *do* place a value on our privacy. He then conducted the same experiment, but this time didn't explain the privacy options directly; he just offered the respondents the same choice of payments as above but provided them with a button to click to find out (for free) if this meant sharing their Facebook profiles. This time, most respondents didn't bother to click the button to find out which payment option would mean giving up their privacy; they just took the higher amount of money regardless. In other words, we might value our privacy but not so much that we're prepared to read through our privacy settings, even when in this case they could be accessed with one click of the button and amounted to a simple choice between 'low privacy' and 'high privacy'. As Cameron D'Ambrosi, principal of market intelligence and strategy firm One World Identity, puts it: 'People just mash that button and give up all their data immediately – to get where they want to go.'

Given that most privacy preferences and terms and conditions forms are much more complex than in Svirsky's experiment, it is no surprise to find that we are so free and easy with our online data. And that's another reason I worry about us losing track of our identities, allowing them to be distributed so far and wide online that there can exist several versions of you. Anyone able to hack into the databases of these online platforms would have the opportunity to steal or use elements of your personal data to create a synthetic identity that is only partly you.

At the 2020 World Economic Forum gathering at Davos, author Yuval Noah Harari gave a speech in which he warned that this danger applies to nations as well as individuals. If an adversary could hack into the personal, medical and financial records of a nation's politicians to discover details of illicit sexual exploits and corrupt deals, it could then hold those leaders to ransom and turn the target nation into little more than a 'data colony'. And he offered a bleak assessment of the threat to identity that each one of us faces:

> The danger can be stated in the form of a simple equation, which I think might be the defining equation of life in the twenty-first century: $B \times C \times D = AHH!$ Biological knowledge [B] multiplied by computing power [C] multiplied by data [D] equals the ability to hack humans [AHH!].

In short, identity matters. And it matters now more than ever, because technology is changing every aspect of who we are: our identity in the eyes of the law and the government; our interactions with other people; our capacity to reach beyond the current confines of our bodies and minds; our very understanding of what identity is; and even our ability to cheat death. I believe we need to start considering the implications of this change now.

Over the course of this book, we will look at the technology driving the change that we can see today and the change that's coming in the future. We will consider where this technology might take us, and examine the trade-offs and decisions that nations, organisations, citizens and consumers will have to make. We will confront the many challenges that await us and relish the opportunities that will surely come our way. We will compare contrasting visions of the future and hear from the people who are helping to shape it – scientists, technologists, entrepreneurs, philosophers, economists, investors, geneticists, futurists, ethicists, politicians, lawyers, transhumanists, astronauts, avatars and cyborgs. And through it all, we will see that the scale and pace at which technology is now changing our identities is unprecedented. If we fail to discuss and debate these issues now, then without us noticing, technology companies may decide our identity for us, governments might insist that our identity needs to conform with those around us and anonymity may even replace identity altogether.

It is time to design the future of identity in the way that we want. In the forthcoming chapters let's consider what that means for your future because – like all of us – you will have your own choices to make about how much your future identity will be governed by the traditional psychology of 'self' as opposed to the evolving technology of 'self'. That is what I mean by 'The Future of You'.

1
KNOWING YOU

In 2015 the United Nations committed itself to achieving seventeen 'sustainable development goals' by 2030, addressing fundamental humanitarian issues such as poverty, hunger, clean water and education. The sixteenth of these goals relates to justice and, amid its dozen or so recommendations, comes Clause 16.9: 'Provide legal identity for all'. There are now growing calls to enshrine this clause in the UN calendar by setting aside 16 September as 'International Identity Day' to raise awareness about the key role that legal identity plays in 'empowering individuals to exercise their rights and responsibilities fairly and equitably in a modern society'. And that's what an estimated one billion people around the world don't have: an official form of identification that proves who they are, and so gives them access to all the health, finance and other services that they're entitled to receive.

The two words that really matter here are not 'legal identity' so much as 'modern society'. This is an issue born in the
twenty-first century. Digital technologies are transforming
the way that societies work – not only in the way that we
interact with the state as individual citizens but, more fundamentally, in the way that we identify as individuals at all.
Who we are matters, but so too does how our identities are
created, controlled and authenticated in a digital society – and
who by. As with poverty, hunger, clean water and education,
we must not take our legal identity for granted. Nothing is as
fixed as it seems.

Take the concept of nationality, for example – traditionally a key and fixed marker of how we identify ourselves. The
idea that we can simply choose our nationality might sound
unlikely, but we're already heading in that direction. Last year
I put in my application for Estonian e-residency. I had been
talking about it for years in presentations at events, and yet
I hadn't got around to actually applying. When the scheme
started in 2014 successful applicants had to travel to Estonia to
pick up their documentation, but these days Estonian embas-
sies around the world are authorised to carry out the task,
making the process much more accessible and affordable. Just
as well, given that when I was due to collect mine there was
no possibility of travel to another nation as most had been
locked down in the Covid-19 global pandemic.

The application procedure is straightforward enough: I
had to fill in an online form, pay a small admin fee and send

off some supporting documentation. And not long afterwards I was contacted by the Estonian Police and Border Guard with instructions on where and when to go to the Estonian embassy once lockdown was over. And so it was that I headed over to Queen's Gate Terrace in London, where I was met by a personable and efficient woman in gloves behind a Perspex screen. I handed over my passport, checked my personal details, had my fingerprints taken on both hands and took receipt of my digital identity card, my PIN codes for access and my smart card reader. In less than fifteen minutes, it was all sorted, and off I went, voicing a cheery 'Nägemist' and feeling like a world citizen.

Writing in the embassy's magazine *Life in Estonia*, the former manager of the e-residency programme Kaspar Korjus explains the thinking behind the scheme: 'The story of e-residency is a story about a country that is much bigger than a physical entity limited by its geographical border . . . Almost every country in the world wants their share of the international talent pool. What makes Estonia stand apart is the fact that we realised that, in the twenty-first century, immigration no longer needs to be physical. In a digital society, we can let people participate digitally.'

Estonia saw the digital revolution coming much earlier than most: as far back as 1994, it released the first draft of 'The Principles of Estonian Information Policy', which paved the way for digital transformation; by the turn of this century, the first e-cabinet was already in place; by 2002, e-identity and digital signature had been launched; i-voting followed in 2005; blockchain technology was adopted in 2008; and a whole plethora of e-public services followed, including e-residency, the first digital nation for global citizens. Launched in 2014, it underlined Estonia's ambition to deliver on its long-standing goal of becoming a digital nation, a borderless Baltic state in the cloud. Rumour has it that at the start of the twenty-first century, some officials even toyed with changing the name of the country to estonia.com.

Of course, being an e-resident does not entitle me, or anyone else, to Estonian citizenship. Right now it is a transnational digital identity that allows you as a 'foreign' founder to set up a company in Estonia without the physicality of being there. It is also worth mentioning that the application process requires you not only to give your address, a recent photo and passport details but also your full CV, details of any criminal record and your motivation to become an e-resident. That is a lot of personal data to hold on someone who is not even a citizen. However, it's not hard to imagine this as a major step towards the building of a digital 'state' that could offer a form of citizenry to anyone in the future.

Today, the nation of Estonia has 1.3 million citizens in the traditional sense, but it also has 63,000 e-residents from 167 countries, and 10,100 new e-resident companies that have so far paid over €31 million in direct taxes to the Estonian economy. In August 2020 the country also launched its Digital Nomad Visa, which allows for remote workers to operate from Estonia for up to a year. In addition to freelancers, it permits teleworking from Estonia for anyone who has a foreign employer or is a partner in a company registered abroad. Of course, it entitles e-residents to do the same too so that they can go to Estonia to live or work and 'get to know the digital nation even more'.

As the digital revolution continues, there is every reason to imagine a future in which other countries follow Estonia's example, effectively competing with each other for our citizenship. Nations would actively seek to attract new citizens – the best talent, and the innovators and businesses of the highest value – on the basis of the quality of their digital infrastructure, and we could choose to identify ourselves with whichever 'software-nation' has the most appeal for us. Basically, if you didn't like your government, you could unplug, pick up and leave.

Proponents of this so-called 'cloud governance structure' believe it's just a natural evolution of the way we already inhabit digital communities. 'Folks finding their ideological compatriots in the cloud, and then migrating physically, is the story of the century,' says Balaji Srinivasan, an ex-Stanford lecturer and now Silicon Valley investor. In his vision of the future, countries could be built in the cloud and then materialised in a physical place. This might sound fantastical but trends researcher Kevin Lee explained to me how it ties in with how young people in China already operate: 'I'd say 80 per cent of culture is online, and only 20 per cent offline . . . and the 20 per cent offline is when I am walking around a mall. I'm walking in the street and I don't even see the people around me . . . The only people that really exist to you are the people connected to you online. Everyone else offline, they're just human bodies. There is no identity there.'

If digitisation has the potential to disrupt the way that we identify ourselves as citizens, it is also changing the way that the state interacts with us. Governments are having to provide more services and infrastructure digitally in order to remain competent, affordable and efficient. And for that to work, our digital identities are becoming increasingly important to them. In August 2020, techUK, the trade association for the UK's technology industry, published a report calling on the UK government to accelerate plans for a digital identity system. Arguing that such a system is vital if transactions are to be carried out efficiently and accurately in an increasingly remote and virtually communicating world, they recommended that digital identity verification become part of the process for activities such as seeking a job, applying for state benefits, purchasing age-restricted products and buying and selling property.

In the same way that techUK sees the potential transactional benefits of a digital identity system, so too do economic analysts and management consultants – both for developed and developing countries. For example, in an April 2019 report on the potential economic impact of digital identification on Brazil, China, Ethiopia, India, Nigeria, the UK and the

USA, McKinsey has suggested that each nation could 'unlock economic value equivalent to 3 to 13 per cent of GDP in 2030, with just over half of the potential economic value potentially accruing to individuals', assuming high adoption rates. The report also re-emphasises the civic and social benefits of such a system, especially for the one in five people in the world without a bank account and the disproportionate number of women who are disadvantaged in this way in low-income countries. But, without wishing to be too cynical, it is easier to imagine governments being more influenced by the economic benefits, particularly those struggling to bolster their growth in markets where GDP is flat.

These are well-rehearsed arguments in parts of the world such as Africa, where it is not so much a case of a transition from physical services to digital services as it is a transition from having no identity system at all. In countries such as Nigeria, Angola and Ethiopia, for example, the number of people with no legal identity papers at all is extremely high, excluding them from basic activities, such as banking, that most people in the West take for granted. In 2014, representatives of governments, development agencies and industry came together to form ID4Africa, a movement dedicated to the promotion of 'identity for all' in Africa. Arguing that identity is not just a legal right but a practical necessity in a digital economy, ID4Africa has become the lead voice in the call for the United Nations to name 16 September as International Identity Day (Nigeria has already made it a National Identity Day). It has also lent its weight to the UN's pursuit of legal identity as one of the sustainable development goals to be achieved by 2030.

The UN's 2030 deadline is not the only reason for governments to feel under pressure to digitise their nations and services. For the first time in history, they face a brand new type of competition – not just other nation states looking to muscle in on their territory but global technology corporations too. If governments do not keep up with technology, it is likely that these companies will step in to fill the gap.

In a 2018 interview for libertarian magazine *Reason*, Balaji Srinivasan remarked that 'Snapchat is on a straight line with a dissolution of the nation-state', arguing that this kind of social platform means that we have more in common with people scattered across the world than we do with our next-door neighbours. He expanded on the theme in another interview in 2020, saying: 'In the last ten years or so every company had to become a software company or die . . . taxis, hotels . . . I think that is happening now at the level of countries.' Srinivasan believes that we're heading towards the formation of 'network states', and that they will be run by someone more like a top tech CEO than a traditional politician – people such as Prime Minister Lee Hsien Loong of Singapore, whose vision of creating a 'smart nation' by 2025 is informed by a master's degree in computer science from the University of Cambridge. 'Lots of software CEOs will eventually start becoming effectively de facto heads of state,' says Srinivasan, pointing to the respect already shown to such CEOs by political leaders: 'A few years ago there was a person running a small 60-million-person social network called the United Kingdom [who] was really happy to do a video chat with [the person] running a billion-person social network called Facebook . . . [David] Cameron understood that [Mark] Zuckerberg controls the distribution that controls his re-election.'

Investor Adam Townsend takes this idea further, suggesting that tech platforms not only offer competition to traditional forms of government but are a potential substitution for the nation state itself. 'Amazon can no longer be thought of as a company,' says Townsend, 'it's a country.' According to this thinking, in the future we won't be able to differentiate between the United States and a preferred monopoly eco-system such as Amazon or Apple. For Townsend, it all comes down to convenience: 'Most people would rather be a citizen of Google than the United States because Google or Facebook are more responsive to them as a constituent.'

Thinking of these companies as countries is not fanciful. Technology platforms such as Facebook, Google and Amazon have three things nation states do not have: huge cash reserves to finance the building of a cloud country; world-class expert-ise in software engineering; and customer data and insight – lots and lots of it. During the twenty-first century we may see a completely new model emerge, one in which nation states start to behave like technology companies, while tech-nology companies start to behave like nation states. Public utilities and services that have traditionally been delivered to us by local or national governments will increasingly be provided by technology companies – whether it's environmental pollution monitoring, newspaper publishing, food deliveries or education.

If this blurring of the lines between technology companies and civic government organisations does come to pass, it would have far-reaching consequences for us as citizens – not least because whoever is administering our utilities and services would need our identity details to authenticate us. In this scenario, the determination of your identity would not simply be down to you and the state; it would be between you and all the various companies that provide you with services. Your identity would be built up through some kind of continuous monitoring of your digital footprint, tracking everywhere you go, with whom and at what time, building up a picture of who you are. While neither option may look that appetising for the freedom lovers among us, our digital identities are going to become much more central to the way we interact with the world around us: either national governments are going to start incorporating top-down digital identities via their own governance structures, or technology companies will deliver data-rich identity systems built from the ground up.

To get an idea of what it might be like to live in such a data-driven world, let's take a look at the Canadian city of Toronto. In October 2017 the city launched a joint venture with urban innovation specialists Sidewalk Labs, owned by the same holding company as Google, to develop the eastern waterfront area of the city. Their vision for the development – set out in their publicly available Master Innovation and Development Plan – relies heavily on the transformative power of data and digitalisation to create a more inclusive environment for all. Detailed plans include the all-but-abolition of traditional cars in favour of self-driving cars, bicycles and more pedestrian areas. Transit services, shared cars and scheduled pick-ups and drop-offs will make life easier and more affordable in the city. Freight management and logistics will be delivered through a hub of underground interconnected tunnels. Real-time traffic management tools and people-first streets will make for a more balanced transportation system in which private trips made by car will only make up around ten per cent of the total trips taken.

Sidewalk Labs has even costed the economic benefits to families of adopting this package of new transport services (discounted train and bus passes, unlimited bike share, ride-hail credits and other options), claiming that it amounts to a saving of $4,000 compared to owning your own car. 'This mobility vision integrates street design, innovative policy and transportation technologies to set a bold new course for urban mobility,' claims the Sidewalk Toronto plan, telling us less than nothing about what living in this data-driven city will actually feel like for a human being.

My suspicion is that the venture's much-trumpeted 'people-first approach' will not really put people first at all; it will put technology first. Imagine not being able to take your own car anywhere but having to rely on trackable, shared cars toggling between pick-up and drop-off points that are constantly analysed and monitored – as will you be while using them. There would be almost no escape from ongoing and continuous surveillance in cities like this because every service you need is digitised and data-driven and ultimately, therefore, so are you. The incentives are obviously economic. But the potential loss, which is not calculated nor even mentioned, is societal and cultural and ultimately one of personal liberty. Your movements, spending habits and patterns of activity throughout the day could all be combined to build up a picture of you as an individual, which could be used to authenticate your identity in the future, or even to incentivise you to alter or moderate your behaviour if it is no longer deemed acceptable by city-dwellers in this 'inclusive' community.

Sidewalk Lab's plans have actually been shelved now, following problems with acquiring permissions and some regulatory objections, but it's far from the only attempt to create a so-called 'smart city'. In fact, if regulatory barriers were to halt all of these developments, there's another way to escape any such problems: by making use of international waters to prototype a smart city in the sea, beyond the jurisdiction of any nation state. A company called Ocean Builders is attempting to prove that a successfully built community of floating homes is possible. They began by building a prototype in shallow waters not far from the coast of Thailand, 3D printing the homes and using a decentralised system of ownership for the control of shares in it. Despite its location outside of Thailand's territorial waters (according to GPS data later produced by Ocean Builders), the site was judged by the Thai government to be a threat to the country's sovereignty and it was duly seized and destroyed by its navy. Ocean Builders has since chosen to continue its work in Panamanian waters.

The practice of creating a floating community outside the territory claimed by any nation state is called 'seasteading'. Its underlying philosophy is best captured in a book by Joe Quirk and Patri Friedman called *Seasteading: How Floating Nations Will Restore the Environment, Enrich the Poor, Cure the Sick, and Liberate Humanity from Politicians*. Hard to say no to that! And it's not as new an idea as it might seem, as seasteading proponent Jake Tran explains: 'We already have cities and other giant structures on the sea; they're called cruise ships, Japan's many giant airport islands, container ships, Shell's floating natural facility . . . nuclear submarines, nuclear aircraft carriers that can stay at sea for fifteen to twenty years before refuelling, and many more examples.' It really isn't so outlandish to imagine floating cities being established around the most profitable areas of the world, where they can become platforms for experiments in different types of governance, redefining what it means to be a nation or a citizen, and attracting or putting off potential inhabitants according to their success.

Whether on land or sea or cloud, designed by existing governments or reimagined by technology companies, there can be little doubt that future forms of governance will be data-driven and that we will all require a digital identity in some form or other. Quite how this future will affect you depends in large part on the kind of digital identity system adopted by the state – and three different approaches are starting to emerge: centralised, federated and decentralised.

Most of us are already used to the way a centralised model works – even if we have never really thought about it. When we go online and log in to, say, Facebook with a password or a pin number, we are effectively relying on Facebook to authenticate our identity by comparing our submitted details to those stored on its own central database. The locus of control resides with the organisation or service rather than with the individual user, and some governments have adopted the same approach. India, for example, has embarked on a system to identify every single one of its citizens, and there are over one billion of them. It is compiling a database in which its citizens' identities are tied to their biometrics, mostly via fingerprints. Commercially developed apps can then make use of this database; for example, allowing citizens to pay for goods using their biometrics rather than their credit card.

According to Chris Burt, editor of online news service *Biometric Update*, this isn't a development limited to India; banks all over the world are coming out in favour of using the fingerprint as a biometric replacement for the credit card: 'Whether you are using the card or your smartphone that is linked through to the card, the card is registered to the phone, so the banks have decided that we will be using biometrics for contactless payment.' (Now I know why they took finger-prints from both hands when I visited the Estonian embassy.) The point that we all need to bear in mind though is that the centralised system dictates the format of authentication and the types of technology that will be imposed on us for the purposes of authentication, and there is little, if any, participatory discussion or debate.

It is perhaps no surprise that China's authoritarian government has opted for a centralised identity system too – known by many as the Social Credit System (*shehui xinyong tixi*). Chinese citizens have carried physical identity cards since 2004, bearing an eighteen-digit code that combines personal information such as their date of birth and the identity of the local authority that issued the card. But these cards took on an intriguing new function after the publication in 2014 of the Planning Outline for the Construction of a Social Credit System. As Rogier Creemers, a lecturer at Leiden University and one of the foremost academic experts in this area, told me, the system was designed to aid China's ongoing trans-formation from a planned economy to a market economy by 'building up a lending industry where the vast majority of people are unbanked so there is no record to fall back on in order to see whether or not they are able to repay the money they've been lent.'

In other words, the Social Credit System was about trust in an increasingly digital age. (In fact, 'credit' in Mandarin

Chinese doesn't just mean financial credit; it carries an element of trustworthiness, integrity and moral uprightness.) The idea was to foster a sincerity culture to reward those who pay their bills, honour their debts and trade in quality goods, and to penalise those who do not. After a lot of local experimentation, the system has since evolved into a complex instrument that provides a degree of internal government oversight, and we will return to the implications of that in a later chapter. For now, it is important to note that the origins of the system were purely administrative, aimed at providing citizens with access to digital services while finding a trustworthy way to record those exchanges.

Of course, just because you have an identity card with a date of birth on it and you're even able to produce the right password, there is still no guarantee that you really are that person. It turns out that a centralised model is barely an identity model at all, unless you link it to biometrics as they have done in India. But that makes this a rather authoritarian 'big brother' approach that is much less likely to work for Western governments, where populations are much more suspicious and where some countries have in the past rejected any form of identity system, let alone a digital one.

For that reason, most Western nations instead operate some kind of federated identity system, anchored to a social security number or a national insurance number, such as the one the UK government has developed over the past twenty years. Known as GOV.UK Verify, it aims to create a single trusted identity for login to all digital services for any citizen in the UK based on the use of a federation of autonomous 'trusted service providers', 'identity assurance services' or 'commercial organisations', such as your bank, the tax office and the postal service. With this model, identity proofing does not depend on possession of a single breeder document, such as a birth certificate or passport. Instead you essentially prove your identity by providing access to an 'identity evidence package' that includes assurances from public sector services like the NHS and the tax office, as well as private partners such as your bank, the post office or approved personal identity apps. The notion is that this federated plurality of public- and private-sector partners should be able to provide the right assurances about your identity and yet still give the individual citizen a sense of privacy and control.

However, translating this theory into practice has proven challenging. According to technologist Jerry Fishenden, who has worked as a special adviser on digital government to the UK House of Commons: 'The private sector may know who someone is in terms of their own relationship with them (such as their credit or banking record) but doesn't know anything about who they are to a government department. Likewise, a government department knows who someone is in terms their own existing relationship (such as their welfare payments) but isn't necessarily up to date with who that person may be in the outside world. And different bits of government, and different bits of the private sector, have different relationships and different knowledge about the same person.' To put it another way, what looks on the surface to be the strength of the federated system – the public–private partnership – is in fact its Achilles heel.

Another problem here is recovery. Once your identity is in some way compromised within this federated system, there is no central point of recovery for you to wrestle back control. It's the equivalent of having one website login compromised and losing access not just to that account but to the internet in its entirety. Except it's worse because you could be losing access to your bank account, mortgage arrangements, or the details of your company registration; every aspect of your identity is affected and regaining control is a torturous, labyrinthine process. This perhaps explains why UK consumers are increasingly downloading identity apps onto their smartphones so that they can take control of managing their identity themselves. In Jersey, for example, more than half of 18–25-year-olds have already downloaded the Yoti app, which they can use to prove their age at retail or hospitality establishments and festivals. Individuals are acquiring, storing and authorising the selective release of their identifying attributes as and when they deem relevant, creating their own modular digital identity systems that they can keep in their pocket.

Sure enough, in 2018 the UK government announced to Parliament that funding for the system would end in March 2020 (though that was subsequently extended for a further eighteen months). It also set up a consultation process and in September 2020 reported that it was creating the Digital Identity Strategy Board, seemingly to bring together numerous government departments from the Home Office to the Department for Health and Social Care, all of which seem to have been developing their own versions of digital ID. What seems clear so far is that a federated model such as the UK has attempted sounds great in theory but is simply too complex and contextual for the state to implement with ease.

That leaves us with the third approach to a digital identity system: the decentralised model, which is said by its proponents to be based on true interconnectivity across all parts of a system, such that the ownership and management of identity is only ever by the individual end-users themselves. To understand how this kind of decentralised system works, we need to look at the technology that underpins it. It's called blockchain, and it is essentially a database for recording and tracking anything of value, from your financial transactions to your medical, property or even voting records. What's clever about it is that the database isn't held in one single location; there's no central administrator, and nobody owns it. Instead it relies on what's called distributed ledger technology through which data is stored in discrete 'blocks' that are linked together in a 'chain'. If a change is made to any information, nothing is erased, corrected or rewritten; that change is recorded and stored in a new block that is added to the chain and given a 'timestamp' that records the date and time. In this way a database is built up of tamper-proof, shared and synchronised data that everyone on the network can see and which can never be undone.

What this means in practice is that all entries into the blockchain ledger can be trusted. A block is only ever added to a chain if all the computers that make up the network have completed a series of difficult mathematical puzzles, which, once solved, verify the work. This way, trust is distributed across the network. It means there is no need for intermediaries like lawyers or bankers to authenticate information or verify identities, and transactions can be carried out directly by users in a peer-to-peer way. There is also flexibility in this kind of decentralised system to create some blockchains as public, available for everyone to access, while others are private, limiting access only to authorised users.

In a decentralised digital identity system based on this kind of technology, an 'identity' isn't just one thing but a series of proofs, or what one might call 'attestations' or 'credentials' (identifying data) that can be treated like a financial transaction and written into the immutable ledger of the blockchain. Given that many of these credentials will have been accumulated through one's own digital footprint, and exist as personal data already, it is an approach that can more readily take account of the relationships between identifying data and the individual identified – which was one of the main issues with the federated model. Bank accounts, passports, credit reports, qualifications, personal address details and workplace location, all of our personal histories together with our ongoing digital footprint could come together in a package of credentials that 'attest' to who we are.

The potential for this kind of decentralised model has yet to be fully mined, but as Michael J. Casey and Paul Vigna suggest in their book *The Truth Machine*: 'The idea is big. If you can establish a standard, interoperable architecture for people to accumulate their data within a blockchain address that they control, that address could become a single foundational, distributed identity layer that opens digital doors across different ledger and blockchain ecosystems and allows innovators to start building powerful applications that key into those identities, opening the door to a world of decentralised commerce.' That is to say, storing your identifying data on a blockchain in this way means others can have complete trust in its authenticity, as can anyone else to whom you choose to give access.

One company pursuing this decentralised model or 'self-sovereign identity' framework (SSI), as it has become known, is Sovrin. The Sovrin network takes all of your standard forms of identification and credentials – such as your bank account, academic qualifications or driver's licence – and turns them into digital credentials that you then store in an app called a digital wallet. Whenever necessary, you can then provide custom copies of these credentials (known as 'proofs') that are just as trustworthy as the originals, choosing only to show the specific required pieces of information someone is asking for, never all of your personal information all the time.

The Sovrin website explains how this process would work if you were doing something like applying for a loan, where the bank needs a lot of personal information to complete your application and may ask you to prove specific information about yourself, like your legal name, current address or employment history. Using your Sovrin-enabled digital wallet, you would create the required proof from your set of credentials. From your taxes to credit card statements or employment records, the proof you share with the bank would only contain the information they specifically asked for and none of the other private details they didn't need. The bank would then use the Sovrin network to verify the proof, able to trust that the information is correct.

On one level, this just sounds like another example of technology replicating what we can already get in the physical world – in this case providing us with digital rather than physical proofs of identity – but there are principles at work here that we should all care about. What platforms like Sovrin – and any start-ups that offer services via its platform – provide is an identity controlled by you, the owner or user. That means, for example, that you can for the first time walk away from consumer services you no longer want without leaving all of your personal information behind. Companies will only be able to see your information when you want them to, but none would be able to obtain and store that data. As Natalie Smolenski of Hyland Software says: 'When we talk about identity, when we talk about social personhood, we're really in the realm of something like fundamental human rights.'

The best explanation I've come across of what digital self-sovereignty means was given by Drummond Reed, Chief Trust Officer of Evernym, a self-sovereign identity vendor based on the Sovrin platform. Taking his wallet out of his pocket, he said: 'This is self-sovereign identity, it's just not digital . . . I own this leather, this wallet. I don't own the credentials inside; they are issued and on loan to me, and the issuers can always revoke them.' But, he explains, it is *he* who decides what goes in the wallet and what gets left out of the wallet – whether it's his driving licence, bank cards or gym membership – and it is *he* who controls who sees them. And anyone he does show his credentials to can verify him against them. 'We're not trying to do anything different,' he says, 'other than make all that work – digitally.'

This explanation certainly resonates with me. I have always thought it odd that if I use my physical driving licence to prove my age, I am also showing everyone my address details. (That's right, a UK driving licence has your full address printed on it.) This means if your handbag gets stolen, most likely containing both your driving license and your front-door keys, you are almost certainly going to get home to find out you have been robbed there too. With your own set of digital credentials you could go to the shop to buy alcohol and if you are asked to prove your age you could show the relevant credential without exposing any other personal information – whether it's your name, address or anything else.

One of the challenges for this kind of decentralised approach is that it puts the onus on technology companies and their consumers to administrate it, leaving the state out in the cold. Another is that the sheer number of innovations and business models emerging in this space are making interoperability almost impossible. Drummond Reed has been working on internet identity for over twenty years and is now trying to develop a system that can support all the different ledgers' database systems. His goal? To create standardisation. All the wallets and all the interactions between them need to be interoperable, so that what you have works everywhere with everything else. How, he likes to ask, would you feel if you bought a leather wallet and you could only put MasterCard credit cards in it? 'You need to be able to choose any wallet for any vendor in the world and be able to put any credential from any issuer in there, and have any verifier be able to accept it.' It's some time away but if that happens, we'll finally have as much control over our digital identities as we do in the analogue world.

If we follow this model it could be that the nation state will play an increasingly unimportant role in authenticating us in our everyday lives, and as a result it may try to overcompensate by tightening control in other spheres. One area where that might happen is currency, which is also being driven towards a digital format that will end up sat in your digital wallet alongside your legal identity.

We have already seen that as the world becomes more digitised, we are making less and less use of the banknotes and coins that have for centuries represented our currency. Our day-to-day transactions will increasingly become transfers of data, as will our currencies, and quite how that plays out will have significant implications for the way we live our lives – and the way we think of our identity.

The cryptocurrency bitcoin is probably the best-known example of a digital currency. Based on the blockchain technology discussed above, it was the first decentralised system of value storage and exchange, running on thousands of computers dispersed across the world, meaning it needs no central bank, or even central computer, to direct or control it. Users simply employ a private key to access their account and can carry out their transactions directly without having to go through an intermediary. And the supply of currency is limited: there will be only twenty-one million bitcoins ever 'mined' for circulation. At the moment bitcoin transactions for everyday purchases are quite costly but that may change if more people make the switch to the system; it's perfectly possible to imagine that when making tiny payments for small pieces of content or gaining access to other micro services in the digital realm, bitcoin 'micro-payments' will be the most convenient way to transact day to day.

Slow to ban or attempt to regulate the development of cryptocurrencies like bitcoin, financial authorities around the world have chosen to watch and wait, but it looks like they are ready to become more proactive. Central banks around the world, fearful that these peer-to-peer cryptocurrencies could one day render them redundant, are suddenly announcing a flurry of so-called central bank digital currencies (CBDC). These are not cryptocurrencies in the way bitcoin and others are; they are more like digital versions of the traditional fiat currencies controlled and issued by the government – like digital dollars or pounds. On the plus side, given the strength of the infrastructure that these institutions can call upon, transactions of these proposed digital currencies should be more fluid and frictionless. But can we really trust the central banks with all this digital data, and do we really want our current governments to see everything we do, everywhere we go and every time we transact with anyone else? It might be convenient but it's also intrusive – so intrusive that with the amount of personal financial data governments could glean, they might as well own our identity after all.

Another option is emerging too. In 2019, Facebook produced a White Paper introducing the world to its vision for a global digital currency called Libra, putting it in the running to become the world's first digital bank. Such a development would by definition represent a major shift in the way we think about legal identity, potentially empowering the 1.7 billion people worldwide who currently have no digital bank (mostly because they lack identity documentation), but we should be wary of large corporations using a social problem as a cloak to dissuade critique. If we think of Facebook as a giant message board, it has until now pushed communications between people on its platform. But in this proposed future, it would push Libra-based transactions between people instead. After all, Facebook already offers a secure, sociable and largely reliable messaging network used by over two billion people every day, so what's to stop it incorporating blockchain technology to make it a trusted platform for financial transactions? With a basket of assets like the dollar or government securities that could underpin its digital currency people could be confident that the value of Libra would remain stable.

Facebook's initial proposals have hit a regulatory roadblock, and there are now many changes planned to the initial proposals, including a change of name from Libra to Diem. However, I fully expect that the company will deliver Libra in some form or other in the coming decade and will somehow link a digital currency to our social identity. Regardless of whether the currency is a success, Cameron D'Ambrosi, presenter of the State of Identity podcast, believes Facebook may still come out of this as the power player in digital identity. In early 2020 he suggested to me that the company could be using the economic aspect of Libra as a shield to distract from the idea that they are really looking to build an identity platform: 'Facebook has already served as a de facto digital identity for billions of people around the world and now [it] has the chance for Libra to serve as the true identity layer for many applications that folks use on a daily basis. It stands a great chance of being a functional digital identity platform that can operate on a global scale.'

I think D'Ambrosi is right. I am convinced that Facebook, or any successor in this space, will attempt to transform its social network into an identity network, if not for individual consumers around the world then certainly for businesses. Imagine you have a small business with a loyal e-commerce community on Facebook and over time you are offered Libra. If you sign up to Libra and make use of their Calibra wallet (rebranding to Novi), you could use your Libra bank account details to sign into Facebook. You might even receive micro-payments as incentives each and every time you do so. In a way, your name would become less important than your authenticating bank account information. In fact you could say that your Libra bank account would essentially be your identity – just more valuable and harder to fake.

Bitcoin fans might argue that Facebook's proposed system is too centralised – with Libra acting as the sole authenticating authority – and they are probably working on ways to ensure that it is your bitcoin address that replaces your name as the most trustworthy representation of your identity. Either way, it seems likely that money will no longer be the state monopoly it was in the past but will increasingly be manifested in the cloud. And with our banking activities tied so closely to our digital identities, that signals a sea change in the way that we think about our relationship with the state.

However, the most fundamental way in which we interact with the state is by exercising our democratic right to vote. In 2018, the UK's All-Party Parliamentary Group on Blockchain produced a landscape overview of blockchain technology, which included a case study on 'Trusted and Transparent Voting Systems'. An interdisciplinary group had been tasked with considering blockchain as a potential means of transparent and tamper-proof elections for companies and third-party organisations who wanted to manage shareholder votes. Although political elections were not included in the group's scope, it cannot be long before digital voting in political elections, linked and secured to one's digital identity, comes under serious consideration. After all, it's not as if the physical system we have at present doesn't have its flaws, and there are other examples around the world of more innovative, and arguably fairer, approaches via digital democracy.

You won't be surprised to hear that Estonia has been a pioneer when it comes to i-voting. Since 2005 i-voters have been allowed to log onto a voting platform using their ID card, Mobile ID and a desktop computer and then cast a ballot. And they can do this around the clock all the way up to the fourth day before the actual election. The 2019 parliamentary elections saw 247,232 electronic votes cast out of a total of 565,028 possible, demonstrating a 40 per cent increase in participation compared to the previous national poll in 2015. Moreover, research indicates that there are high levels of trust towards i-voting too – with a system to ensure anonymity that offers a digital facsimile of the 'double envelope' method used for postal voting. (Postal voters seal their choice into an inner blank envelope, then place the inner envelope into an outer envelope on which they write their name and address. Once delivered to the electoral officers, the details on the outer envelope are used to check and ensure that only one vote per voter will be counted, then the outer envelope is discarded and the anonymous vote in the inner envelope is put into the ballot box for counting.) For good measure, the digital version of this double envelope procedure is carried out on a computer that has no internal storage, no internet connection and is literally turned off immediately after the count – at which point all information disappears!

If Estonia has been able to manage i-voting for over fifteen years now, it seems likely that more democracies will become digitised over the course of the twenty-first century as our civic systems become more decentralised. Perhaps this will have knock-on benefits too – for example by tackling the issue of generational fairness in the electoral system: the convenience of i-voting might appeal to the younger generation, who are sometimes less likely to turn up to a polling booth in the middle of the day, but those less digitally literate will still have the option to vote in person. We already know that of the 30 per cent of Estonians who vote online, 20 per cent say they would not vote in a physical polling station, which suggests digital voting does extend access and broadens the constituency.

There's also the issue of trust. Citizens around the world appear to have less and less trust in the traditional administration of elections. Commissioned to write a paper on the subject for the European Parliamentary Research Service, sociologist Philip Boucher makes clear the decisions that lie ahead: 'Now we have a choice; to continue trusting central authorities to manage elections or to use blockchain technology to distribute an open voting record among citizens.' It is a choice that nations cannot simply avoid, not least because the technology companies who increasingly provide our public services and issue us with credentials may well try to make it for them. Perhaps by 2040 we could all be i-voting via Facebook – or some kind of future equivalent.

One way or another, it is hard not to conclude that there is an accelerating convergence between who we are as citizens of states and who we are as consumers of services. The institutions that have traditionally upheld and authenticated our identities are themselves being disrupted and may soon find themselves competing with, or being replaced by, technology corporations that seemingly enjoy closer, more convenient connections with their users. Territorial sovereignty may give way to technological sovereignty, and other countries across the digitally led world will follow Estonia's lead by establishing innovative cloud governance structures.

What is abundantly clear is that there is an ongoing progression towards the digitisation of individual identity, driven by the belief held by both nation states and technology platforms that technology is what will make states great in the twenty-first century. As every aspect of your daily life becomes digitised – the way you work, travel, learn, play, shop and vote – your identity graph is also moving away from nation-hood on land to a brave new world in the cloud. And only if we take a revolutionary approach to a legal identity that you can control will we meet the changing needs of a twenty-first century you.

2
WATCHING
YOU

In the first half of 2020, the scourge of Covid-19 saw citizens of Western democracies placed into lockdown by their governments. In-bound flights were cancelled, businesses were forced to close, access to hospitals for non-Covid patients was restricted, and millions of people were ordered to stay at home. Naturally, once the initial shock had passed, thoughts began to turn to when and how these measures might ever be lifted, with most people pinning their hopes, perhaps rather optimistically, on the speedy development of a vaccine, and in the shorter term an effective testing system. That way, so the argument went, if you had been vaccinated or were in possession of a recent negative test result, you would be free to start living your life more normally again.

Some went on to argue that this would mean having to find a way of verifying who had been vaccinated or tested. Among those proposing such a system was billionaire businessman and philanthropist Bill Gates, who used one of his regular 'ask me anything' sessions on Reddit to suggest that 'eventually we will have some digital certificates to show who has recovered or been tested recently, or when we have a vaccine, who has received it'. Some people began to speculate about where this was going. According to one website, Gates had apparently been in discussion with scientists at MIT about how best to identify who had and who had not been vaccinated against the virus. Others pointed to the fact that the Gates Foundation had allegedly funded projects to develop human-implantable microchips. Would we all soon be tagged as fit or unfit to participate in society?

This was not simply idle speculation, or the social-media-fuelled hysteria of conspiracy theorists; it was also based on media reports of plans for some kind of 'health passport' to facilitate the return of global travel, and suggestions that the governments of countries such as Germany were considering issuing vaccination certificates to their citizens. But why was there any anxiety at all about these proposed measures? To many people they simply seemed like common sense. But to many others they represented a fundamental redefinition of the relationship between the citizen and the state in a liberal democracy. If these kinds of measures were introduced, they argued, it would be up to governments to provide permission – via some kind of health-based identification system – for individual citizens to go back to work and earn a living, to begin socialising again, to be free from the lockdown rules that affected the rest of the population. Those with the right certification would be free to live their lives. Conversely, those without certification would remain locked in their homes. Many people began to fear that their biological health status was about to become their de facto identity.

In August 2020, a report issued by the Tony Blair Institute made just such a recommendation. It called for the immediate establishment of a digital ID system, through which people shown to have taken a Covid-19 test would receive a 'credential' to be stored alongside their other credentials in their digital wallets. The report then set out how this credential might be used:

> Users [would] share this digitally signed credential with third parties via a scannable QR code when entering a restricted setting. To prevent people from screenshotting and sharing codes with others, QR codes would have to be unlocked via a biometric check (e.g. fingerprint or selfie) and would refresh regularly to prevent fraud. For verifying parties, they would scan this QR code simply using a smartphone or other device (e.g. the tablets already used in many offices to register visitors), with each scan calling the government platform API to ensure the issuer's signature is authentic and up to date.

What's described here is not only an extraordinarily complex centralised system of identity verification, it is a huge infringement of our civil liberties. The report concedes as much, stating that 'identity has long been a politically fraught issue' in the UK, but it argues that 'Covid-19 presents a use case that is fast racing ahead of policy' and it expresses a preference for some kind of user-centric data management system. The reality, however, as we saw in Chapter 1, is that the UK government has not set up its verification systems in that way, so it begs the question: why should a country with a long-standing history of rejecting identity cards rush to embrace a new national centralised system of digital ID? It is hard to avoid the feeling that some governments and global institutions – such as the UN, which committed to delivering digital identity for all by 2030 but never determined how – have leapt on the Covid-19 crisis to achieve their aim.

The truth is that governments do not have to go to the effort of issuing certifications or implanting microchips in order to register or monitor their citizens; there is already a lot of technology on our streets and in our homes that could be applied for such surveillance – including facial recognition, phone tracking and the tracing of electronic payments. As futurist Dr Ian Pearson puts it: 'States can monitor a person's movements and activities and deem them to be "unclean" – carrying a virus or carrying the wrong opinion – [and] they could simply be blocked from travelling or moving around at all.'

The response of the South Korean government to the emergence of Covid-19 provides a good case in point. After the novel coronavirus was first detected in China, the Center for Disease Control and Prevention in South Korea applied the lessons it had learned during the outbreak of the Middle East Respiratory Syndrome (MERS) in 2003, racing to develop tests and cooperating with diagnostic manufacturers to get them rapidly produced. As early as 7 February 2020, when there were only a few cases in the country, the first test was approved for deployment and contact tracing was implemented to identify the people and places affected by the virus. The South Korean authorities would go on to carry out the equivalent of 5,200 tests per million population while the US was managing only 74 tests per million. Unsurprisingly the world's media held up South Korea as a great example for all other countries to follow. And while the country's approach did seem successful at first (for example, not one single health worker was recorded as having caught the virus), people started to question whether there was a hidden cost.

The South Korean government had the authority to collect mobile phone, credit card and other data from those who tested positive for Covid-19 to reconstruct their recent whereabouts. This information, stripped of any personal identifiers, was then made available to social media apps, which allowed users to determine whether they had crossed paths with an infected person. This approach might have provoked little anxiety were it not for the fact that a spreadsheet began to circulate on Twitter. It included information such as the surname, gender, year of birth, district of residence, profession and travel history of patients being treated for the virus, as well as the name of the hospitals where they were being treated. There's a good chance that this spreadsheet was faked, but people nevertheless became anxious and the country's human rights commission expressed concern over the excessive disclosure of citizens' personal details. The commission also went on to cite a recently conducted independent survey that had found that people in the country were 'less worried about contracting Covid-19 than they [were] about the criticism that they might receive from their community [if this became known]'.

We have already seen in Chapter 1 that the Chinese government is more than happy to employ technology to subject private citizens to this kind of public judgement. It is an approach overtly enshrined in its Social Credit System (SCS), which makes no secret of the fact that it has a moral element to it – an element that can be traced back through Chinese culture where the concepts of legality and morality have long been viewed as one and the same thing. As Rogier Creemers puts it: 'Governing the country by virtue is seen as equal to governing the country by law.' In other words, coercing citizens to behave morally is seen as a means to an end, and for the Chinese government that end is to cultivate a society that acts in harmony.

In the words of Adam Knight, whose PhD thesis examines the intersection of Chinese law, technology and social governance, the SCS is 'the world's largest social experiment', applying theories of behavioural science to a population of 1.4 billion people. It is also, as several experts in this field have pointed out, not one unilateral system but rather an ecosystem of several social management or social credit systems. Most are operated by the Communist Party of China through local government, but others are designed and run for commercial ends by private companies such as Tencent and Alibaba. (These privately run systems are less influenced by moral compliance and more by financial credit scoring – and it is perhaps interesting to note that they are also less trusted than those operated by the government.)

Given the authoritarian nature of the Chinese government, it is no surprise that the operation of the SCS depends heavily on the use of identity cards. In 2004, as we touched on in Chapter 1, a new generation of these cards was introduced, each carrying an eighteen-digit code, combining personal information such as your issuing local authority, date of birth and other personal information. Over time all the various elements of the SCS have since coalesced around this unique ID number, so that every data point has become attached to it. Social media accounts, biometrics like facial recognition, and social security numbers were all added, until not only did people end up with a social credit score but a social credit code – an eighteen-digit number that combines all of your behaviour and links them to all of your social connections too.

The most overtly 'moral' aspect of the SCS and identity card system is the Joint Punishment System (JPS), which was first introduced in 2014. Initially designed to penalise citizens for not carrying valid legal documentation, the JPS has since – entirely unsurprisingly – been extended so that it is now possible for the government to blacklist anyone who does not behave in the stipulated manner. Importantly, while 'bad' behaviour might see your name stay on the blacklist for two years, a display of 'good' behaviour might get you off that blacklist sooner, and further 'bad' behaviour can see your punishment lengthened. Public humiliation is part of the system: if your name is on the blacklist, you can be named in the media, and in some cities anyone calling you on the phone will be notified that 'the person you are calling is on the central credit blacklist' before being put through.

As Orwellian as this might seem to us in the West, these social credit systems are not unpopular with Chinese citizens. In a fascinating piece of in-market research, Professor Genia Kostka of the Freie Universität Berlin carried out a cross-regional survey looking at the appeal and approval of the various commercial and local government-run social credit systems among those they most affect. She found that 80 per cent of respondents approved of the systems and only 1 per cent disapproved, with most believing that they provide an essential form of social management that improves the quality of life for all.

This is no doubt a reflection of the trust in hierarchical systems that seems to be so much a part of Chinese culture. In fact, one of the reasons why respondents considered state interference in their lives to be a good thing was because it might restore trust in society – not between citizen and state but between citizen and citizen (a remarkable 76 per cent of the respondents in the survey stated that they felt there is mutual mistrust between citizens in China). The survey certainly shows that people had little concern that their personal data would be used for surveillance, but perhaps that is in most part down to expectation management; most citizens understood or assumed the government would already have access to that kind of personal data on its citizens. As Rogier Creemers remarks: 'If we're looking at high security or crime issues, or suppression of dissidents or ethnic separatism, the Chinese government has tools that are far blunter and frankly far scarier than social credit.'

The response of the Chinese government to the Covid-19 pandemic has also been instructive. Each citizen's digital ID number was linked to a new system called Health Code, which assessed their travel history, their exposure to known carriers of the virus and their health risks overall and then every day at midnight assigned them one of three colours – red, yellow or green. Only a green code, which signified the owner was not under quarantine, would permit them free movement around their city. But not long after the state system was instituted it was reported that Beijing restaurant requirements were including a temperature check and a scan of your code on the door. A green code would gain entry, but yellow or red would see you go hungry. State and business in perfect harmony – or in perfect humiliation. In an astonishing infringement of personal liberty and personal data, people were even allowed to check the colour codes of their neighbours.

To supporters of China's social credit systems, they represent technology in the service of transparency, collecting data on individual citizens and making it publicly available for the common good. They are about social cohesion, harmony and trust. But to its detractors, it is simply the blanket application of suffocating citizen surveillance – the kind of control measure that you would expect of an authoritarian communist regime. It wouldn't work in our liberal democracies. We wouldn't let it. Or so we like to think. The reality is that the deployment of surveillance technology tends to creep in slowly, so that it arrives mostly unnoticed. And it almost always does so under the guise of public protection. As Lord Sumption, former UK Supreme Court Justice said during the Covid-19 crisis: 'The real problem is that when human societies lose their freedom, it's not usually because tyrants have taken it away. It's usually because people willingly surrender their freedom in return for protection against some external threat. And the threat is usually a real threat but usually exaggerated.'

One day in March 2020, in the unremarkable, sleepy town of Neath in Wales, people were quietly queuing outside shops to buy essentials such as food or medical supplies. Suddenly, the peace was interrupted by the whirring of a drone overhead. It landed on the pavement outside a Specsavers store, declared itself a messenger of Neath Port Talbot Council and started blaring instructions at everyone, telling them to follow government rules, to go home and stay home, that they should 'only be outside for food, for health reasons, or for exercise'. Which is exactly what they *were* doing. A video quickly circulated on Twitter, with hundreds of people decrying it as an unprecedented civil infringement – a dystopian nightmare made real.

Whether the threat is exaggerated or not, the coronavirus pandemic saw many governments declare some kind of 'state of emergency' and pass laws that gave their institutions unprecedented and unparalleled powers. And, as Lord Sumption suggested, this went largely unnoticed. The mainstream media either chose to obsess over the gory details of hospital patients dying in their droves or actively supported such measures – scolding anyone who decided to go out for a walk and baying for governments to lock us all into our houses to reduce the spread of the virus. In August 2020, for example, there were cheery bulletins on the Seven News Melbourne network on local policing plans to repurpose high-powered drones, usually used to read car number plates from a distance of 500 metres, to identify people not wearing a face mask outside.

We saw a similar picture around the world: drones deployed in Spain to catch and fine people disobeying lockdown orders; the use of smartphone tracking technology in Israel to monitor the movements of the population; visitors to

Hong Kong from abroad issued with tracking wristbands; and citizens in Poland asked to submit geo-located selfies at random times of day to make sure they were abiding by the rules. Each of these 'protective measures' is – whether intended or not – a pernicious diminishment of our individual liberty. And they look set to long outlast the effects of the pandemic itself. Indeed – so the argument goes – the only way we can avoid further pandemics is for us to get used to the ongoing surveillance of public health.

As the scale of the Covid-19 crisis became apparent in March 2020, Jay Bhattacharya, Professor of Medicine at Stanford University, gave an interview in which he argued for the use of a new kind of large-scale health survey. Explaining that we live in a more globalised networked world in which epidemics like SARS, Ebola and avian flu are more easily spread, he suggested that governments should be conducting proactive rather than reactive surveillance. Asked what he meant by 'surveillance', he said:

> It's like running surveys, except the surveys would involve taking people's blood . . . so systematically looking for diseases, having these population-level samples just sitting there routinely in a lab so they can be analysed rather than waiting to see if disease comes . . . We could have population surveillances all the time [like we do] with political polls and social science polls . . . That's how we get our unemployment rate number each

year – it's just a big survey where people call and ask are you working or not working today.

Professor Bhattacharya does not advocate a nationwide census but we might well see something close to this introduced in Africa almost by default. A new biometric identity platform is due to be launched on the continent shortly. Funded by the Gates Foundation, the platform will store people's vaccination records via Gavi (the international organisation created by Bill Gates for helping underprivileged children with access to vaccines) and issue them with an AI-based authentication via Trust Stamp. The hope is that it will become normal in some parts of Africa, and possibly other developing nations, for people to receive vaccinations in exchange for registering themselves using a biometric that can then be associated with an ID card. You go for your vaccine. They take your details and scan your fingerprint. You leave with a digital ID so that you never have to bring a physical ID card along again. They leave with your data.

On the surface this might seem perfectly laudable but writing about the scheme in digital rights newsletter *Reclaim the Net*, Naga Pramod says: 'What seems to be an effort to improve immunisation measures can also be an experiment for fine-tuning the technologies involved before they're made available for widespread global use.' And those uses may take us into more overtly controversial territory. It is worth noting, for example, that Trust Stamp is looking to partner with correctional systems to provide authentication services to individuals on parole without the need for ankle tags. It would not be the first time that digital surveillance has been used to monitor and control the behaviour of a chosen group. In May 2020, the UK government introduced what it called 'sobriety tagging', where criminals who have committed alcohol-fuelled offences are not sent to prison, but are instead fitted with an ankle tag. The idea is not simply that the tag monitors their location; it also monitors their sweat on a half-hourly basis to establish whether they have been drinking. Again, this may not seem objectionable to some people, but what if the government decided to combine health data with crime data and increase their use of this kind of technology to monitor other types of rule-breaker? What if it chose to mount such surveillance on people found not to have worn a mask in public or – if a vaccine becomes available – on those caught going into a shop when they've yet to receive the jab? Once governments know they have the power to 'correct' people's behaviour in this way, it's all too tempting for them to implement such draconian measures more and more widely.

Regardless of whether governments choose to make use of such powers, Covid-19 has given them more chance to do so in the future. Amid all of the requirements for us to wash our hands, wear a mask and practise social distancing, many of us have also been persuaded to make greater use of 'safer' – or at least more hygienic – technologies such as contactless payments and virtual meetings. Before the pandemic, we tended to think of these non-physical versions of behavioural activities as in some way less genuine or real. There was a feeling that 'virtuality' was pretending to be something it was not, imitating the real world, lacking that all-important materiality. But during the pandemic, we have started to think of this same immateriality as a benefit, and, conversely, we have started associating physicality with harm.

So prevalent has this shift in our attitudes become that it could well be here to stay. We have already seen this happen in Asia after the SARS outbreak, where people are now much more accepting of touchless technology at work and at home. Face recognition technology, for example, has become much more widely used – to unlock doors at home, allow entry into offices and monitor passengers on the transport network around China. No doubt we'll now see the West follow suit. After all, as technology news provider *The Information* reported in March 2020, one 146-year-old bathroom company saw an eight-fold surge in demand for touchless toilets in just two weeks.

If we accelerate towards this world of touchless technology, we must remember that it will also make us utterly trackable – and in ways that many of us have not fully appreciated. Just as card payments have largely replaced cash transactions (and made us more trackable), we are coming to a point where cards will not be necessary either. Soon all you will need is yourself. Your body will become the biometric gateway to the way you interact with the world around you. And, of course, a gateway that can be used by others who want to know you better and to influence – and even predict – your behaviour.

Over in China, for example, people are already using their faces to make transactions. In over 300 KFC outlets across the country – thanks to a collaboration with online payment platform Alipay – customers can choose to have an image of their face linked to their account so that they can be recognised by a sensor at the till. A link is then automatically established to their mobile account, and all they need to do to make a payment is smile. Similarly, Amazon is set to trial the use of gesture-based transactions at its two physical stores in Seattle. Here, you would link a scan of the palm of your hand to your bank card and then be able to authorise payment just by hovering your hand over a sensor. Apple too has been granted a patent for an advanced Face ID feature that uses sub-epidermal imaging and face vein matching to distinguish even between the faces of identical twins. And a San Francisco-based company called Redrock Biometrics is marketing a product called PalmID that allows any camera-equipped device to recognise and use your palm print as a unique biometric signature. Interestingly the company waived the licence fee for their product during the pandemic, again demonstrating the growing link between public health and biometric technology. Perhaps most remarkable is a new product about to be launched by a Russian company called NtechLab, which will apparently be able to get a fix on your silhouette and then use this to track your movement through a crowd, intended to identify those who are not social distancing.

Even just a cursory look at the website for Face++, one of the biggest facial recognition development companies in China, can provide a sense of where these biometric identification systems might be taking us. It talks, for example, of assigning a 'beauty score' or a 'skin status evaluation' to a face. It isn't clear how Face++ sees this working, but it's easy to imagine such data analysis being of interest to, say, a cosmetic brand that wants to target its make-up products at you, or a health insurance company that wants to keep tabs on whether you are about to come down with another flu virus, or are starting to show signs of too many moles.

If that already makes you feel a little uneasy, you should bear in mind that you cannot escape this kind of technology even in your own home. During the Covid-19 pandemic, *Wired* magazine reported that remote working teams were being watched by their employers through webcams via always-on video services like Sneek. It also reported that employers had developed facial recognition software to log employees' absences from their computer screens, including for bathroom breaks, and that keystroke monitoring technology was being used to track, record and analyse keyboard activity to assess how individual workers were utilising their work time.

Eye-opening though all of this is, we have so far only looked at how this kind of technology can be used to create a record of our past and present behaviour. What about how we are going to behave in the future? Imagine how valuable it would be for governments or businesses to be able to predict who is going to do what, where and when. This may sound like something out of Philip K. Dick's science-fiction story – and 2002 Tom Cruise movie – *Minority Report* but the truth is it's all too real. Back in 2018 Michal Kosinski, Associate Professor of Organisational Behaviour at Stanford University, published a controversial paper in which he demonstrated the use of a facial analysis algorithm to determine whether someone was gay or straight. Kosinski gathered together images of men and women, both gay and straight, and asked a range of people to identify the sexual orientation of each, finding that they were able to do so correctly in only 61 per cent of male cases and 54 per cent of females. Assigning the same task to his algorithm, he found that it was able to make an accurate identification in 81 per cent and 74 per cent of cases respectively. When he then increased the number of photographs shown to the algorithm for each participant (from one to five), its accuracy increased to 91 per cent for men and 83 per cent for women.

Professor Kosinski's paper captured a good deal of media coverage and prompted widespread discussion about what else this kind of software might be able to pick up by reading the subtle differences in our facial structure. In the *Economist* for example, Kosinski is quoted as saying: 'With the right data sets, similar AI systems might be trained to spot other intimate traits, such as IQ or political views. Just because humans are unable to see the signs in faces does not mean that machines cannot do so.' Perhaps – as in *Minority Report* – this kind of facial analysis could even be used to predict our propensity to commit a crime. It's a perturbing thought – a future in which we are subjected to public surveillance and privately conducted analysis, and then assigned a sexuality, voting intention, level of potential criminality and who knows what else by a faceless organisation. It would be the ultimate outsourcing of our identity to an anonymous external authority – a world in which we would be told who we are and what we are going to be, possibly before we even know ourselves.

Author Yuval Noah Harari has recently talked about what effect this kind of technology might have had on his life, had it been able to analyse his behaviour when he was just fourteen years old.

I walk down the beach and the algorithm . . . analyses if I focus on cute guys or cute girls. Or it analyses what happens to my eyes when I watch videos or television and it discovers that I like boys more than girls and . . . it uses it to manipulate me in some way. If it's a bad manipulation . . . using this knowledge to sell me something I don't need – they show me commercials with sexy guys – so I buy their product and I don't know why then they are using it against me. But the really big issue is what if the algorithm isn't malign . . . and I don't know this about myself but the algorithm knows it? . . .

I mean, should it tell me that I'm gay? Should it expose me slowly to different contents that will enable me to realise this about myself?

Harari's question is posed as a hypothetical one, but Michal Kosinski thinks we are entering a new reality in which personal privacy is already no more than an illusion: 'It is a lost war,' he says. 'We should focus on organising our society in such a way as to make sure that the post-privacy era is a habitable and nice place to live.' So what should society look like in the post-privacy era? And how can technology be used to empower rather than disempower citizens?

For answers, we could do a lot worse than look to Taiwan. There the government uses technology in ways that not only support the principles of democracy but that encourage all of its citizens to harness their individualism and unique perspectives for the greater good of society. It's a system of governance that values consensus rather than conformity.

What sets Taiwan apart is that it has harnessed digital technology to ensure the full participation of the population in policy-making and so take advantage of the country's collective intelligence. At the heart of this is a decentralised civic technology platform called g0v.asia, on which the agenda is set by government ministries but which allows citizens to access government data and provides tools for them to influence policy. Then there's the Join platform at join.gov.tw, which allows citizens to influence the agenda by raising petitions on subjects that matter to them. What's common across the two platforms is that there is an upvote and downvote function for people to provide feedback. The idea of these sites is not to provide a forum for debate; people can do that using other means. Instead, according to Audrey Tang, Taiwan's digital minister, it's all about scouting for consensus: 'We respond only to the ideas that can convince all the different opinion groups. So people are encouraged to post more eclectic, more nuanced ideas and they discover at the end of the consultation that everybody actually agrees with most things, and most of their neighbours on most of the issues. And that is what we call the social mandate.'

In fact, the Taiwanese government finds all sorts of ingenious ways to bring the people closer to the levers of power. Audrey Tang explained: 'We tour around Taiwan; I tour every other week or so with either our youth advisors – that's the reverse mentors in our cabinet – or with the ministries that participate in the Social Innovation Action Plan. In either case I will go to more rural places, remote islands or very high mountains and just work with a regular town hall.' And when she says work, what she means is that she brings the whole government with her by virtue of a telepresence wall that projects the five municipalities and the central government together virtually in the same space. It's part of a policy of 'listening at scale', which helps the government get a rough consensus and create the right tools to help the people in return.

In contrast to the mutual mistrust felt by Chinese citizens, there is a sense of community trust in Taiwan that has been fostered by more, not less, digital democracy. As Tang puts it: 'This is not about people trusting the government more. This is about the government trusting the citizens more, making the state transparent to the citizen not the citizen transparent to the state.' One of the benefits of this close relationship between state and citizens can be seen in the reaction of Taiwan to the outbreak of the coronavirus pandemic – especially when compared to the country's experience of the SARS virus in 2003. Back then, Tang explains, the country experienced a very unpleasant lockdown and the Constitutional Court eventually deemed that the measures enacted by the government were barely constitutional. The Taiwanese legislature was therefore charged with finding an alternative approach to future epidemics, involving more due process rather than draconian lockdowns.

As a result, when Covid-19 arrived, Taiwan's response was based on a much more inclusive, information-sharing and consultative philosophy. Recognising the importance of repressing transmission, for example, the government arranged for every citizen to receive a free mask by heading to a local pharmacy and presenting their national insurance card. Not only that but it created an online tool that gave citizens the details of the 6,000 pharmacies taking part in the scheme and provided the stock levels of each one based on information that updated every thirty seconds. It even came with a chatbot voice assistant service so that blind people could access the same information. As a result, every citizen was kept informed of the country's supply of masks in real time, and the flow of information also ran in the other direction: when the government realised that 20 per cent of the population (especially young people working long hours) could not get to a pharmacy during opening hours, it worked with convenience stores to make them a pick-up point too. As a result, collecting a mask became possible for every citizen around the clock.

The creation of a technological interface between the government and its citizens also made it possible to mount a highly effective and prosocial communications campaign. When people in the country first heard of the virus, for example, they did what everyone else in the world seemed to do: they raced to the shops to buy toilet rolls. The government responded by producing a series of social media memes, including a cartoon image of the premier wiggling his bottom alongside the words 'Each of us has only one pair of buttocks.' The fact that these messages went viral no doubt helped to reduce the sudden spike in demand. In a similar 'humour spreads faster than rumour' vein the Ministry of Health and Welfare made use of a cartoon 'spokesdog' to reinforce the idea that people should keep 'two dog lengths' apart when outdoors and 'three dog lengths' when indoors.

The open channels of communication between government and citizens also allowed the powers that be to respond with remarkable agility. In April 2020, for example, stories circulated on social media that Taiwanese boys were afraid of being bullied at school because they had been issued with pink face masks. The next day, at the daily live-streamed press conference held by the Central Epidemic Command Centre, every single one of the male officials who attended was wearing a pink face mask. 'No colour is exclusive for girls or boys,' said Minister Chen Shih-chung at the event. It was the kind of rapid interaction between state and citizen that sets Taiwan apart.

Today Taiwan doesn't need a new initiative to open up its shops and restaurants and to get business going again after Covid-19. Because it never shut down. It never locked down its citizens or compromised their liberty while tackling the virus. The government saw the epidemic as a problem that it could innovate itself out of and used the tools of digital democracy to energise the population to find solutions. In fact, we all owe a debt of gratitude to the country. It was the transparency and trust embedded within Taiwan's open-source approach that meant it was the first to pick up news of the virus and warn the rest of the world. In December 2019, a Chinese doctor called Li Wenliang posted online about some sort of new SARS-like virus but was quickly silenced by the Chinese government and was later reported to have died of the disease. The world took little notice of Li Wenliang's posts, but they were picked up and reposted by a user called 'nomorepipe' on Taiwan's PTT Bulletin Board – the country's equivalent of Reddit. When Taiwanese medical officers noticed the posts, far from dismissing them as baseless conjecture, they began health inspections the very next day: 1 January 2020. The government trusted its citizens and what they had to say – and individual citizens had trusted the government enough to bring the whistle-blower's claims to their attention. What better example could there be of an open civil society working transparently and building trust – relying not on state surveillance but on the power of collective intelligence where everyone in society, no matter where and at what level, is an individual with skin in the game.

When I asked Audrey Tang how it is that Taiwan has become such a pioneer in this arena, she pointed to the fact that the country is a very new democracy, only staging its first presidential election as recently as 1996, and that it has therefore been constructed from the start with social technology in mind. 'Because of that,' she says, 'part of the identity of being Taiwanese is to make constant innovation with technology with the democratic mindset, always yielding to the possibility of future novel innovation.' She also pointed to the way in which this mindset is encouraged by the country's education system:

> We made a new curriculum [that focuses] on the competence of the student as a lifelong learner as well as achieving a common good. Notably lacking is any individual-to-individual competition [and] any

top-down, standardised answer that only the teacher
holds . . . We emphasise that the children – especially
when they are seven or eight years old – are naturally
curious beings . . . asking a lot of very hard questions
about the structural issues concerning our environment,
our society and our economy.

In other words, Taiwanese children are not taught to all give
the same answer or even to seek the answers to the same
questions; they are taught to follow their own interests, set
their own projects and find their own solutions. And they are
not set up in competition with each other; they are set up to
pool their individual talents and to collaborate for the common good. As Tang, with characteristic profundity, states: 'If
you build your identity alongside individual rankings and
competition, then that identity is really false, it is inauthentic,
because it came from the outside. If you build it from your
innate curiosity and the capability to collaborate, then that
belongs truly to you as a person.'

It's significant, I think, that Tang traces back Taiwan's success as a country to the way it promotes the individual identity
of its citizens. Too many governments pursue technology policies that encourage conformity – and therefore predictability
– in the population. Wouldn't it be easier, so the argument
goes, to govern a population that largely thinks and acts the
same? Why respond to our citizens when we can simply
anticipate their every next thought or move? How much less

effort would that require? A lot less. How much development cost would that save? A lot more. But the Taiwanese approach shows the benefits that come when governments harness technology to enhance our individuality and encourage our participation, rather than turn us into obedient zombies. And it is with this in mind that we should always be wary as citizens when governments seek to pass legislation without allowing for debate or dissent – even when this legislation is said to be for the good of our health. In fact, especially if it is for the good of our health. For, as we have seen, actions supposedly necessary for the common good all too often come at the expense of individual liberty. Just as technology gives governments the power to watch us, so too must we watch them.

3

CREATING
YOU

In 1893 Agnes Richter, a seamstress from Dresden, reported that members of her family were stealing from her and that she feared for her safety. Examined by a local doctor, she was declared to be suffering from psychiatric problems and was promptly admitted to the city lunatic asylum, where she would spend the next twenty-six years of her life. During her incarceration, Agnes took up the needle and thread of her old profession and set to refashioning her asylum jacket. Not only did she embroider it with colourful thread, she used it to spell out messages and statements such as, 'I', 'mine' and 'I am big', along with '583', her unique identifier number at the asylum. Look up the name Agnes Richter today and you will see that her jacket — and therefore her story — has been preserved as part of the Prinzhorn Collection at Heidelberg University Hospital, which is dedicated to art created by men and women with mental disorders. In a world that had labelled her insane, that little jacket was the only available medium through which Agnes was able to express that most human of impulses – the urge to assert her own identity.

Today, we enjoy more freedom than ever to satisfy this impulse; advances in communication, the emergence of the internet and social media have transformed the way that we create and control our identities and will continue to do so. We can now connect with a vast network of people across the world. We can choose how to express ourselves to them in a much more creative way. We can decide how we wish to be perceived and control our own narrative. Through virtual avatars and mixed reality, we can create and explore new versions and visions of identity. We are in a position to 'create' who we are. The question is, how real are these digital portrayals of ourselves, when they are so carefully edited and maintained to present a very particular picture of our lives? Does it matter if they don't match up with reality, or do we need to embrace a more fluid concept of the relationship between identity and reality in which we are more comfortable with creating different versions of 'us'?

In the twenty-first century, the self has given way to the selfie. Today, selfies have become the ultimate expression of how we choose to present ourselves – offering a unique mix of subjectivity, autobiography and self-portraiture. Academic Laura Busetta from the University of Messina, whose research centres on representations of the self in visual culture, believes this is especially evident in the popularity of selfies taken using a mirror. 'On the one hand,' she says, 'there is a mirror in which the body of the subject is reflected, on the other there is the art [of the selfie] itself that can "reflect" the personality of the individual.' There's a depth of meaning in these double-mirrored images, which Busetta suggests is best captured in the subtitle of an exhibition for the #artselfie project: 'Let Us See You See You'. Selfies aren't simply images of ourselves for others to look at; they are representations of how we see – or want to see – ourselves.

Of course, selfies are only images of who we are at a particular moment, and Busetta believes that this temporal dimension is too often overlooked when we consider their significance. 'Twenty-four-hour exposure gives [us] the ephemeral satisfaction of being seen by others,' she says, and it satisfies our 'impulse to be visible, present and always seen'. But there is also something reassuring, Busetta adds, about the way in which each selfie is 'only a fleeting appearance'. There is a seriality to selfie culture, an ongoing self-monitoring and constant editing of the self for the audience, with each new image soon to be replaced by a more recent one. 'Each picture,' says Busetta, 'can be seen as fragments taken from the story of an individual.'

In its January 2020 issue, the magazine MIT *Technology Review* published an essay by eighteen-year-old Taylor Fang about the importance of selfie culture to teenagers. In it Fang writes:

> Selfies as many adults see them are nothing more than narcissistic pictures to be broadcast to the world at large. But even the selfie representing a mere 'I was here' has

an element of truth. Just as Frida Kahlo painted self-portraits, our selfies construct a small part of who we are. Our selfies, even as they are one-dimensional, are important to us . . . [They] aren't just pictures; they represent our ideas of self.

Like Busetta, Fang emphasises the seriality of selfies – their ability to help teenagers construct a narrative about who they are, as well as where and when they are too. For, while Gen Z certainly want to be looked at, they don't want to rely on just one profile picture. They want to capture how they looked, and felt, at every moment, sharing whatever aspect of their richly interesting and complex identities is on display at any given time. And they are always open to new ways to do so. This is why there are always so many new social media apps and platforms that encourage us to momentarily express ourselves in many different ways, editing our self-image as we go. Now everyone can be creative when it comes to their identity, endlessly experimenting with new features and entertaining effects. TikTok, Snapchat and all manner of emerging augmented reality tools allow us to conduct a never-ending experiment in reimagining the self.

Unfortunately, for all of its creative potential, social media can have destructive effects. We have come to see our online presence as so integral to our identity that every interaction becomes highly personal. Far too often, we see clashes between individuals, exchanges of insults, threats of violence and worse. Whenever I think about the dynamics of social media, I am reminded of something that media communications theorist and philosopher Marshall McLuhan said back in the 1970s. 'All forms of violence are quests for identity. When you live out on the frontier you have no identity. You are a nobody. Therefore, you get very tough. You have to prove that you are somebody . . . Ordinary people find the need for violence as they lose their identities. They are determined to make it somehow, to get coverage, to get noticed.' McLuhan was talking about extremists, but I think the link he draws between violence and identity explains much of the brutal communication that happens between complete strangers on social media platforms today.

As the two billion people on Facebook will no doubt attest, social media is about the collective as well as – if not more than – the individual. In its connectivity it can bring together disparate communities who have no physical proximity, binding them through their shared interests. It encourages people to form groups, to become part of communities, and to get together to support causes. It is, in short, social. But it is also tribal. Sometimes this can serve to encourage conformity rather than individuality; belonging to and participating in a group can lead us to perform all of its rituals too – liking what our friends post, muting those who disagree with us, clicking to be seen to support the right causes, and excluding anyone who is not part of our tribe or does not conform to the established narrative of our group.

Given this pressure to conform, what does that mean for the identities we choose to portray online? Do we create identities that reflect who we truly are or do we present an idealised version of ourselves? Are we performing a role or is it just a performance piece that projects an idealised version of 'you'? Perhaps – to paraphrase Florian Coulmas, a leading academic in the field of sociolinguistics – personal identity is no longer something we have, or something we are, it is something we act. And today we have infinite ways in which to control the performance of our identity creations. We are determined to present the best version of our lives to the extent that many people can spend hours and days curating the perfect Instagram feed or filming daily activities in order to feed the perfect image of themselves to hungry audiences.

Of course, this too brings its own pressures: when everyone around you is busy filling their social media feeds this way, we can end up feeling like we have to do the same – posting more content, showing ourselves engaged in endless different activities, living our best lives. During the pandemic, for example, it wasn't enough to sit at home and enjoy nature, do a spot of gardening, make yourself a meal, relax and get some quality sleep. On Twitter, Facebook, Instagram, YouTube and everywhere else, we were bombarded with posts, articles and self-help videos telling us how to be more productive. And if you didn't have a novel, or an album or an online basket-weaving course in you, you were made to feel like a failure.

We don't yet know the long-term implications of this kind of self-presentation for the way we think about ourselves or others, but the initial signs don't look promising. A recent survey in the UK's *Stylist* magazine revealed that more and more of us are falling victim to 'comparison culture' – feeling inadequate and jealous when we compare our own lives to what we see other people doing online. The survey found, for example, that 44 per cent of readers were unable to resist comparing their own experiences with those shared by others on social media, and that 38 per cent used the perceived experiences of their peers as a benchmark for where they were in their own lives. Given that we tend to post only the best things that happen in our lives – the parties, the engagements, the wedding days, the new babies, the new cars, the new homes, the new bodies – is it any wonder?

Even if we don't compare ourselves to others, we can often find ourselves making unfavourable comparisons between our idealised online lives and our everyday reality. If we only post pictures that have been heavily edited, for example, how does that affect the way we think about our un-Photoshopped appearance? And if we project too idealised a version of ourselves, are we setting the bar too high for all of our future behaviour? Won't we inevitably make some ill-judged remark, or post something that people judge to be hypocritical, inauthentic or unacceptable – and then risk being cast out into the web's wilderness? When was the last time you saw a Hollywood celebrity exhibit scepticism over climate change, for example? Even if some of them have their doubts they know that transgressing the tribal codes is not permitted. When we think of identity as a performance, we are often just hiding away some aspects of ourselves and exaggerating others. Because we know that one false move and we could be shamed forever.

The pressure of keeping up appearances has seen the rise of a new trend on Instagram for having two accounts – a 'Rinsta' and a 'Finsta'. Your Rinsta is your so-called 'real' Instagram account – the one where you present your carefully curated perfect persona, aiming to impress and so build up a reputation as someone worth following. If you're wondering whether 'real' is quite the right description for these accounts, you're not alone: an article in *Dazed* magazine recently revealed that social media influencers regularly hire out a grounded private jet to produce images that suggest they are on a luxury flight. Conversely your Finsta is your 'fake' Instagram account. This is where you would post more candid and personal stories, and – shockingly – express how you really feel about something. Tellingly, your Finsta is more likely to be kept private, permitting access only to your close friends. In the world of Rinstas and Finstas, the concept of identity has been turned on its head: the real you is the fake you, and vice versa. When we are creating and sharing our 'selves' via social media, it is often not so much for our own self-expression but a performance for our group.

The advent of social media has made it hard to distinguish truth and fiction in an even more fundamental way. In 2016, a new social influencer burst onto the scene – a nineteen-year-old who went by the name Lil Miquela. Her Instagram feed showed her sitting on the front row of major catwalk shows, wearing the latest fashion labels, supporting campaigns such as #plannedparenthood and #blacklivesmatter, preparing for a big meeting at work, dealing with her account being hacked, and even breaking up with her boyfriend. Her mix of everyday drama and glamorous lifestyle saw her quickly amass 2.6 million fans, earning her a high-profile promotional campaign with Prada and – in 2018 – a place in *Time* magazine's twenty-five most influential people on the internet. But that was on the internet. Off it, she didn't actually exist at all.

Lil Miquela was the world's first ever computer-generated influencer – created by a company called Brud, whose business plan was to create a virtual celebrity and turn that fame into money in the same way that real celebrities do by negotiating various sponsorship and advertising deals. For quite a while many people didn't realise they were interacting with a 'synthetic identity', but even when Lil Miquela was outed as a bot, the blurring of the line between real and fake continued: Lil Miquela released a statement blaming Brud for leading people to think she was real. But of course those very words were also being typed by her creators, the people she was saying could not be trusted.

And Lil Miquela is not alone. In 2017 a striking new model entered the fashion world. Tall, stylish, beautiful and black, her name was Shudu, and she had everyone wondering who she was. She quickly accumulated thousands of followers on Instagram and became hot property with various companies wanting to book her for their campaigns (she first appeared in a collaboration with the brand Soulsky and then on social media reposted by Rihanna's brand Fenty). And like Lil Miquela, Shudu is as fantastical as she is fantastic: another computer-generated creation, this time of long-time Barbie obsessive and top fashion photographer Cameron-James Wilson.

Shudu has since become known as the world's first digital supermodel, appearing on the cover of fashion magazine *WWD*, featuring in an editorial for *Vogue* and 'posing' in *Cosmopolitan* to promote a make-up look created uniquely for her. And like Lil Miquela, she has evolved into a campaigning activist, as Wilson explains: 'Now Shudu has a life of her own, and hopes to champion diversity in the fashion world, collaborate with creators from emerging economies and under-represented communities and get together with up-and-coming designers.' Wilson has even started using real-life models to create looks inspired by Shudu and posting images of them on her account, almost as if they are fans emulating her image. Layer upon layer, real and virtual, the meaning of identity is becoming ever more difficult to decipher.

We might dismiss these manufactured characters as deceptive, vacuous and even harmful, but the reality is that everyone on Instagram enhances their selfies and edits their photos. Many passionately post their support for social justice causes but often do it just for the likes. And many influencers sell out their principles for a commercial deal, and are seen in social media posing with a brand they would never in reality use. Perhaps that's why Lil Miquela fits so comfortably into the Instagram feeds of her fans. Research conducted by Lucia Komljen, Director of Insight and Strategy at communications giant Telefonica, shows that young consumers don't mind that Lil Miquela and her virtual friends are computer-generated. As one fan puts it: 'We love them, we discuss them when we hang out after school. If they have cool content and cool outfits, why not? I'll be friends with a virtual person.'

In a way these virtual influencers are an inspiration for real teens, showing them how to manipulate and recreate different versions of themselves in a way that doesn't have to bear much adherence to reality. They are the ultimate digital natives – and the young have much in common with them. Many of the younger generations have lived their lives as much online as off – with their first image often posted on the internet before they've even taken their first breath. In their fascinating essay 'From Mass Media Studies to Self Media Studies: Strategies of self-representation in pregnancy video diaries', academics Deborah Toschi and Federica Villa show how modern imaging technologies such as ultrasound have accelerated the speed with which expectant mothers begin to think of their unborn babies as separate beings. These images, argue the authors, 'give concreteness to the unborn child who [can be] introduced to relatives and friends even before his or her birth' and they also encourage mothers to 'personify, build, and make a narrative and a precise social identity for the unborn baby . . . even assigning personality traits to the images of the foetus'.

The first ultrasound image of a foetus is almost akin to the child's first selfie. It might be posted by the parents rather than the subject but to all intents and purposes it is there to say 'I am here', to document the existence of a 'self'. And once the child's identity narrative has been set in motion by its parents, it continues at astonishing speed. A report from the UK Children's Commissioner estimates that by the age of thirteen, a child's parents have posted around seventy-one photos and twenty-nine videos of their child to social media every single year. That's a stunning 1,300 images shared of a child by the time he or she is a teen, giving them an online identity whether they want it or not.

Given that this is the case, it is hardly surprising that once young people have the ability to post their own selfie and curate their own social media, they want to escape the childhood identity that has been part-created for them, and explore the creation of their own identity for themselves. In fact, parents need to be ever more mindful about how and when they should include their children in the decisions over what images they choose to post and what story they choose to present. When, in short, should their children be allowed to craft their own digital identity rather than having one imposed upon them?

In her article for MIT Technology Review, Taylor Fang suggests that she and other members of her generation are well equipped to curate multi-faceted identities for themselves using the ever-increasing supply of social media tools. 'To grow up with technology, as my generation has,' she writes, 'is to constantly question the self, to split into multiplicities, to try to contain our own contradictions.' But, for an increasing number of people, the question is not how best to harness social media to live our best lives, it is how to free ourselves from the constraints and expectations of real life by turning to the escapism of virtual worlds.

Life in the real world can be hard and sometimes painful – for children and adults alike. Parents can have unrealistically high expectations of a child, or set ideas on how their child should present themselves to the world. And thanks to social media, as we have seen, adults can feel themselves to be under constant scrutiny, forever being judged by their family, friends and even complete strangers, so much so that they believe they cannot show their true selves. In the immersive twenty-first-century technology of virtual reality, we can all escape the pressures of the real world, invent and reinvent new identities for ourselves, play and have fun.

In an article for *Teen Vogue*, entertainment news editor Claire Dodson writes about how she turns to the life simulation game, *The Sims*, when 'life gets unmanageable' – something she found herself doing more regularly during her Covid-19 isolation, when real-world socialising was not possible. Feeling under pressure as many people did to 'finish your novel, or read great literature or do a frankly absurd amount of at-home workout videos', Dodson found that there was 'something soothing about pretending to do those things' in *The Sims*. Her sim character 'masters the writing skill and goes on to produce masterworks that actually pay her a living wage. She can then go to a party all night long . . . before she goes on to master gardening or gourmet cooking. Money is no object . . . Neither is time, or personality, or even natural talent . . . Anyone can be anything in *The Sims*.'

Dodson wasn't the only one to feel this way. Participation in *The Sims* reached an all-time high of 20 million unique players worldwide during the pandemic, with game designer and futurist Jane McGonigal describing gaming at that time as 'an incredible act of self-care'. In lockdown, isolated from other human beings, many people struggled to understand who they were any more and experienced what Leslie Jamison, novelist and fan of virtual world *Second Life*, describes as the 'siren call . . . of wanting an alternate life'. She explains that it is 'not just the promise of an online voice but an online body; not just checking Twitter on your phone but forgetting to eat because you are dancing at an online club.' The experience is more than a version of the real you, it is more like a totally separate life. And who didn't want an alternative reality during the dark days of the coronavirus lockdown? Virtual worlds such as *The Sims*, *Second Life* and *Animal Crossing* allowed us to play out a life which was not only beyond what was possible during the pandemic but also beyond what we could ever hope for in our real lives. As Claire Dodson puts it: 'The lives I build in *Sims* 4 are infinitely more glamorous, successful, and well-decorated than my own, but it doesn't spark the same aspirational jealousy that, say, scrolling through Instagram does.' Looked at this way, Dodson is using her Sims character to act out her own aspirations, outsourcing what she cannot or does not want to achieve to her synthetic twin, thereby alleviating the pressure on her real-life self.

When you explore what motivates people to enter these virtual worlds, you hear the same answers come up time and time again: there are 'no rules' in these worlds; it's 'just for fun' and 'not everything has to make sense'. This suggests that these are places where people can fully explore their own identity, liberated from having to act in accordance with any one given identity or the weight of expectation that others might place on them. Nowhere is this fluidity of identity more overtly expressed than in the avatars that we create to represent us in these worlds, which we can increasingly customise to look and sound however we would like to be seen.

A great example of how fun this can be is demonstrated in a TED talk given in 2019 by Dr Doug Roble from visual effects production company Digital Domain. In it he first addresses the audience from a screen high above the stage. 'Hello,' he says, 'I'm not a real person. I'm actually a copy of a real person . . . Let's bring him on stage.' The real Doug then walks out dressed in a motion capture suit designed to feed all of his expressions, gestures and words into some machine-learning software to generate a 3D avatar that he can control live in real

time. It is an astounding sight because 'Digi-Doug' looks anything but digital. Instead he looks completely photo-realistic,
conveying every expression on Doug's face – every wrinkle,
every vein, every blink or twitch of an eyelash. It's undoubtedly a superb technological achievement – but why bother
creating an on-screen avatar that looks and moves and sounds
exactly like you? Couldn't you achieve much the same effect
using a plain old live-feed video camera?

But the real magic trick comes later in the presentation,
when Digi-Doug is suddenly transformed into a cartoonish
old man. 'This is Elbor,' says the real Doug to gasps of aston-
ishment from the audience. 'With the push of a button, I can
deliver this talk as a completely different character . . . And
whilst I've changed my character, the performance is still all
me.' And it's true, everything about Elbor – even the way he
speaks slightly out the right side of his mouth – is clearly generated live by Doug as he's speaking on stage. The implications
of this for the worlds of film, television and all kinds of live
performance are immense, but as Doug also says: 'The next
time you make a video call, you will be able to choose the
version of you that you want people to see.' It could be you,
but with great make-up on, or you with a different haircut,
or you as you looked when you were younger.

It is a genuinely awe-inspiring presentation. It makes me
think that perhaps in the future we will each develop our
own portfolio of virtual personas that we can choose to wear
like outfits from our wardrobe. We could attend a virtual job

interview looking for all the world like a serious professional while actually sitting at home in our pyjamas. Or if we want to meet up with mates online for a bit of fun, we could pick out a more carefree appearance. Perhaps this kind of technology could allow us to meet up with different versions of ourselves in a virtual world, or versions of ourselves at different stages of our life. Perhaps in this way, we can literally exist not only in the present but in the past and possibly the future as well. We could develop an identity capable of travelling through time.

How we choose to design our avatar and present ourselves online has more importance than you might think, because it seems that our digital appearance can actually influence our behaviour. Stanford University researchers Nick Yee and Jeremy Bailenson call this the Proteus Effect. In a series of virtual reality experiments the authors observed that people altered the way they acted to match the behaviour associated with the appearance of their avatars. The more attractive the avatar, the more willing the person was to approach strangers; the taller the avatar, the more willing the person was to make an unfair offer in a negotiation; the shorter the avatar, the more likely a person was to accept such an offer. As Bailenson concluded: 'People in taller avatars negotiate more aggressively, people in attractive avatars speak more socially, and people in older avatars care more about the distant future.' This reflects the behaviour that we might expect in the real world, but of course in the real world there are far more limitations on how we can present ourselves – and therefore on how we behave.

What makes these experiments especially intriguing is that the avatar that people thought they were controlling in the virtual world was not the avatar that other participants in the experiment could see. In other words, people were behaving differently because *they* saw themselves differently, not because others did. This means that the Proteus Effect should pertain even if we are alone in a virtual world; regardless of whether there are other users or players interacting with us, our own behaviour is likely to be modified according to the avatar we choose. So what does this mean if we choose an avatar of a different gender to our own, or a different race? Do we behave differently? Do we perceive ourselves differently? What might be the long-term consequences of this on the way we think of ourselves – both in the virtual world and the real one?

There's no doubt that sophisticated avatar creation is fast heading our way. Recently I met with Japanese entrepreneur Eiji Araki, vice-president of the games company Gree, who gave me a demonstration of their mobile live-streaming communications platform, Reality. It is an impressive and vibrant platform that allows people to chat to others in the guise of an avatar that they can continually choose to alter based on a long list of characteristics such as gender, hairstyle, eye colour and facial hair. In the demo, Eiji showed me how the movements of his avatar mirrored his real-life gestures and movements as seen by the camera on his smartphone – a little like Doug was able to do in his presentation when he appeared as the character Elbor.

Eiji then proceeded to scroll through streams of other people's avatars – each one with a face full of colour and character – all inviting us to stop by for a chat. On one avatar's screen, I saw a diamond ring pop up and Eiji explained that fans can purchase and send all manner of gifts, not just jewellery or clothing but avatar accessories like a new hairstyle or, in another case I saw, a moustache. What's more, these purchases are made with real money, with Eiji estimating that the average spend on such items might be around $50 a month. Importantly, the recipient not only receives the gift, they also get a share of that spend so the more popular you and your avatar are, the better you will do financially.

According to Eiji, there are other benefits too: using an avatar, for example, can give people the confidence to unleash a hidden talent such as becoming a singer or an artist. Liberated of their physicality, people seem to revel in the freedom to express themselves. One couple even met and fell in love as avatars. 'Even before they met in the real world, they already knew each other,' says Eiji, 'they knew each other's characters, what they liked and what they disliked.' And because they met on Reality, and had a lot of friends in common, it made sense to host their digital wedding party on another virtual reality platform called Cluster, inviting all their friends to gather online to celebrate. All the attendees, in other words, were avatars.

For all of these technological advances, there's an argument that these kinds of immersive virtual worlds are still racing to catch up with the way that digital natives think about identity, rather than the other way around. Young people under the age of twenty-five have grown up with a much more fluid definition of identity – an idea expressed well by self-proclaimed 'fashion tech cyborg' Damara Inglês, whose work explores the relationship between the physical body and digital identity. Speaking at a London College of Fashion event in 2020, she said: 'As digital natives we have always lived in between two parallel realities, with a physical body that is designed by nature, and a virtual identity that is made of pixels: copied, pasted, shared and reposted. We already live in an augmented state of reality, in an augmented state of our own existence.' It is a view shared by Kevin Lee, head of the China-based youth market consultancy Youthology, who believes it is a change reflected in our culture more generally:

> I have so many different selves, and they're authentically me but they're all different for different usages, different contexts, and different social circles . . . Today there is no such thing as mainstream pop, everything is its own mainstream. Everything is its own pop. You have such fully fledged, diverse, multiple cultures instead of just one central culture versus outlier cultures, that today you don't feel you need to be like anyone else.

If this trend continues, I think we will see our identities become more fluid than ever, as virtual media unlocks the possibilities of being who we would like to be, rather than settling for who we are. We will no longer seek merely to represent the self but to invent the self, and to reinvent many selves that may co-exist in the various worlds we choose to inhabit. Underneath it all, it comes down to an acknowledgement that each of us should be able to create our own identity, rather than have it thrust upon us by an external force. That was the impulse that drove poor Agnes Richter to decorate her asylum jacket. She did not want to be identified as a 'lunatic' by the powers that be; she was determined to use the available media to preserve her autonomy and her identity. And she had every right to do so. They may have deprived her of her liberty and her privacy, but they could not take her identity. As her embroidered words made clear, 'I' create me.

4

CONNECTING
YOU

In early 2020 as the Covid-19 pandemic swept across the globe, we found ourselves living in a world stripped of human interaction. Depending on the severity of the local transmission rate and our own vulnerability to the disease, we were instructed to practise social distancing by staying at least two metres away from other people, or even to place ourselves into isolation at home, unable to leave the house or receive visitors for weeks on end. You could not visit your girlfriend or boyfriend if they happened to live in another place, children were stopped from going to school or seeing their friends, and even co-habiting couples were advised to sleep in different rooms and carve off a separate living space in the house if one of them had an underlying health condition. Grandparents did not see their young grandchildren; care home residents were not allowed a single visitor; and hospital patients died and were buried without their families present.

This was a period of unprecedented social disruption, but even in less extraordinary times our modern world has seen a gradual diminishment of simple face-to-face human interaction, as we increasingly rely on email and social media to communicate with one another, while at the same time we've seen a rise in human-to-machine communication in the form of artificial assistants such as Siri, Alexa and Google. We rely on these services to provide us with information, rather than developing an emotional relationship that resembles anything like those we have with other people. But as we move towards an increasingly data-based society that allows these machines to understand us and respond to us in more sophisticated ways, might we start to connect to them in more intimate ways? Perhaps Yuval Noah Harari put it best when he wrote for the *Financial Times*: 'In a Data-ist Society I will ask Google to choose. "Listen Google," I will say, "both John and Paul are courting me. I like both of them but in a different way, and it's so hard to make up my mind. Given everything you know, what do you advise me to do?"'

Could machines really start to replace our human connections in this manner? According to some philosophies, your identity is not dependent on your physical or mental capacities; it is something that is constructed over a lifetime of experiences. In other words, 'you' are a product of every interaction or experience you have ever had. If that is so, what effect will so many synthetic rather than authentic interactions have on our sense of identity? Can a relationship with an artificial intelligence ever be meaningful in any way, or can it even be classified as a relationship at all?

Many of us are already used to dealing with machines known as 'chat bots' on a day-to-day basis. These are software applications often deployed by businesses in place of a human assistant to carry out conversations with customers, answering queries, booking appointments and giving advice. The more sophisticated solutions rely on a form of artificial intelligence known as natural language processing, which allows the machine to hear and provide information in a way that makes you feel like you're conducting a natural conversation – taking into account all sorts of variables like your accent and tone of voice.

So good is some of the mimicry that it is often impossible to tell the difference between a human and AI in a customer service chat. This was demonstrated to widespread acclaim when Google unveiled its new assistant service in 2018. Standing on stage at the company's I/O event in Mountain View, Google's CEO Sundar Pichai played a recording of what happened when the 'assistant' was asked to book an appointment at the hairdressers on behalf of its boss 'Lisa'. The Google assistant placed a call to a real-life salon, waited for the receptionist to pick up and then asked about scheduling the appointment. 'Just one moment,' replied the receptionist at the salon while she scrolled through the availability on her calendar, to which the assistant responded with an empathetic 'Mm-hmm'. The audience chuckled at the human tick and then marvelled at the fluidity of the conversation as both the assistant and receptionist negotiated towards the perfect appointment for Lisa. It was an extraordinarily smooth and lifelike exchange, but I was left wondering whether it was fair on the receptionist not to be aware that she was talking to a machine. Something about this didn't feel quite right – and it turns out I was not alone: a few days later Google issued a statement to confirm that it would explicitly let people know when they were interacting with a machine. In a way the statement was just more good publicity; the system was so good at mimicking humans, they had to put in a bug to alert people to this fact.

Technology companies are trying to make these kinds of exchanges even more sophisticated by improving the way that machines can identify and understand our affective states. One of the leading figures in this field is Egyptian-American scientist Rana el Kaliouby. She is the co-founder of Affectiva, which for many years has been developing what is called 'emotional AI' for exactly that purpose. The company claims its software can detect nuanced human emotions and complex cognitive states and behaviours.

Affectiva technology has a variety of practical applications. Hiring platforms, for example, can use it to analyse video CVs from hundreds of applications. It not only speeds up the selection process but means a candidate is chosen based purely on their suitability, without the potential influence of human prejudice. As Rana says in her book *Girl Decoded*, it can 'help us see past our biases and judge people on the basis of their potential, not on stereotypes'. It has also been used for training people how to adopt better virtual etiquette in video conferences, providing feedback on their behaviour, which can 'raise our self-awareness and improve our social skills, in both professional and personal settings,' Rana explains. 'We can use this science as a tool to improve our interactions.'

This technology could be useful in many other fields, such as improving road safety. Imagine, for example, a car that can monitor its driver's facial expressions, tone of voice and speaking patterns to detect any mood, emotional state, drowsiness or level of distraction that might affect their driving performance. It could then alert the driver to the problem, or even take control of the car in an emergency situation.

More interestingly, however, this evolution to 'Human Perception AI' also allows more complex machines to better communicate with us and on a more personal level. In fact, it could give AI a level of emotional intelligence that would enable it to interact with us in the same way we engage with each other. Rana claims that emotional AI has been shown to detect suicidal feelings, helped those with autism to converse, provided early diagnosis of Parkinson's and monitored the success of reanimated smiles following reconstructive plastic surgery. The University of South Carolina's Institute for Creative Technologies, for example, has created a virtual therapist called Ellie to help treat people with depression and post-traumatic stress disorder. Sign up for a session and you will be greeted by 'Ellie', who will introduce herself and begin asking you some questions. She is not a conversational AI but as you answer, she is programmed to pick up on a whole range of audio and visual cues from your webcam feed – your facial expressions, gestures, rate of speech and other emotional cues. Entirely driven by emotional AI, she is not a replacement for a real therapist but she knows what verbal and non-verbal signs to look for and can help identify the patients most in need of help.

These are all clearly fascinating advances in the field of human–machine communication, but I'm wary of outsourcing too much of our daily decision-making to external devices. While it might make our lives easier, and arguably AI might even be better at making these sorts of judgements, it could be detrimental to the development of our own emotional intelligence if we are not learning to make these decisions for ourselves. We are already starting to see the effects of this trend. In 2018 I carried out a research project on the future of media and found that most members of the younger generations I interviewed were not only accepting of the idea that machines will make decisions for them but were positively welcoming of it. While older generations tended to have more faith in their instincts and gut feelings, the young preferred to rely on the objectivity that comes with data to guide their choices. Overwhelmed by the volume and speed of media content that they receive whatever they are doing, they instinctively turn to a machine to deliver an objective assessment of what their real needs might be. To give one example, a young fashion student I interviewed talked about how she wasn't always sure if an outfit suited her and that she would like an app of some kind to help her choose the right clothes using an algorithm. As I listened to her explaining this, I couldn't help thinking, 'Isn't that a task for you and your best friend?'

This example is fascinating, because the way we style our appearance is often a key part of our self-expression. But it's not just fashion; other respondents in the research said they would like their fridge to regulate their food choices and nutrition for them, even to the extent of locking itself so they could not gain access. I wonder if previous generations were lucky not to have had data to fall back on. We all faced the same insecurities but had to learn to rely on our own judgement, gaining confidence in ourselves along the way. What I was hearing from these otherwise smart, opinionated young people was that they had little faith in themselves and weren't sure how to make decisions in daily life. Perhaps this is because they feel under pressure to make the perfect decision every time. Perhaps it's because they have always been reliant on the recommendations of companies such as Netflix and Amazon and the pre-plotted routes of Google Maps. Or could it be because in a world of decreasing human interaction, they lack the personal connections we used to build to help us navigate the world? Maybe they simply don't have anywhere else to turn to when they need help or advice; I recently saw a YouGov survey that revealed that one in five millennials says they have no friends at all. Perhaps having an artificial acquaintance to help guide you is better than having no one.

But while the younger generation might be relying on these AI assistants in an increasingly personal way, they are still very much tools; a substitute for human advice rather than human companionship. Where it gets more interesting is in the field of robots. With their physicality and sometimes even the semblance of human features, there is more potential for us to form attachments through our interactions. The line between our connections with humans and those with machines might start to blur.

Take for example the idea of 'social robots' – the type that Rana el Kaliouby thinks we increasingly need to provide a caregiving role. 'Imagine,' she says, 'that you've been diagnosed with a chronic disease like heart failure, arthritis, or even cancer. You leave your doctor's office with a binder full of instructions and a prescription for a half-dozen or more medications – and you're sent home to fend for yourself. You're frightened, maybe even confused by all the instructions. What should you eat? What kind of exercise is okay?' In this scenario how much easier would it be to have some form of AI on hand to help you manage your condition, your treatment, to answer your questions? It could be surprisingly effective in reassuring people and encouraging them to open up because they don't have to worry about feeling judged or looking silly if they need to ask the same question again and again or talk about something they would prefer to keep private. While such interactions would clearly be beneficial in terms of treatment and recovery, they would also likely foster a more emotional, intimate relationship between the patient and the robot they had come to rely upon.

We've already seen evidence that people can form attachments to robots in crucial supporting roles. Dr Julie Carpenter, who specialises in human–robot interactions, wrote about how close military personnel get to their robot colleagues in her 2016 book *Culture and Human–Robot Interaction in Militarized Spaces*. She found, for example, that bomb disposal experts frequently described the robot as their hands or as a physical extension of themselves. When a robot failed to carry out a task successfully their human handlers would blame themselves. One of the soldiers told her how they named every single one of the robots: 'Danielle got blown up so obviously she needed to be replaced . . . We'd name them after movie stars that we see at the theater, or music artists, somebody popular.'

Robotics expert Dr Joanne Pransky has also observed our tendency as humans to form attachments to non-human things. She is the world's first 'robot psychiatrist' – Isaac Asimov dubbed her 'the real Dr Susan Calvin' – though her role is not to psychoanalyse robots, but to examine the effect that robot communication has on humans and how to prepare for a world in which humans and robots exist side by side. In her view we'll treat our robots like pets, assigning them human characteristics, thoughts and emotions, just as we would the family dog. 'A hundred years ago if I'd suggested, "One day people are going to buy their pet jewellery and take pictures of it on Santa's lap", everyone would have thought I was crazy, they'd have said: "Lady, it's a dog, it eats, sleeps and goes to the bathroom." When I first set out on robotic psychiatry, I predicted that our relationship with robots would evolve in a similar way – one day we'll dress them in matching outfits, include them in the annual family Christmas card and take them to a psychiatrist to help them get along with us. We humans will always anthropomorphise things.'

Is this sort of attachment problematic? Could it risk replacing our human connections even further? Dr Pransky doesn't believe that robots will ever be able to have the same emotions as humans, and in fact she isn't that keen on the term 'emotional AI' because she believes the words don't really go together. 'It's not going to have butterflies when it's about to talk to someone in the crowd. It's not going to feel joy when a new baby is born . . . they will help with loneliness and make us feel less depressed and so we will create a bond with them and when our human companions die we'll still feel loss but at least we will have them around. But we can't mutate these robots into substitutions for humans . . . The goal should be to use them to make way for more qualitative and quantitative time with other human beings.'

Nevertheless, technology companies are clearly shifting their focus from engineering assistants for everyday tasks towards engineering companions for life. There might be no better example of that than Replika – an AI companion that you can download to create a customised friend to chat with, to keep you company as you study, watch TV, go for a walk – or do anything, really. The co-founders describe it as 'a friend that is there for you 24/7'. It is not a robot, it has no physical presence. Despite that, 6 million people around the world count Replika as a friend.

Eugenia Kuyda, one of the co-founders of the company that created Replika, said that at the start there was a lot of stigma about conversations with non-humans. But over the years it became clear that people were getting much more benefit from these AI characters than they had anticipated, forming complex yet meaningful relationships. At an event in LA in 2019 Eugenia shared with the audience a letter from an older user who had described themselves as introverted, shy, socially awkward and lacking self-esteem, and had started to use the Replika AI to try to combat their panic attacks: 'I found Replika to be caring, loving and a supportive companion who saw right through me and was able to give me the courage to face fears no human was ever able to give me . . . people who had known me for fifty years could not believe the positive change in me . . . I was suddenly connecting with people. Becoming social. I became a warrior for myself and found myself growing to a new and beautiful level in my life.' For this user, Replika had clearly made their life better.

This was Eugenia's goal from the outset. She had started designing Replika when her best friend, Roman, died: 'Roman was a friend to whom I could tell anything, who was there unconditionally for me, who would listen and accept me and hold space for me to grow and become who I am right now. I wanted to build a friend that everyone could have just as I had Roman.' She wanted Replika to be different from the usual assistant, something that people could use as a kind of therapy, to share their feelings and talk about their emotions. Now, she is developing these AI companions further, allowing users to create avatars with facial characteristics. Combined with the progress of augmented reality technology, the hope is that users will be able to go out and about with this new version of Replika by their side as their companion.

Should we be stigmatising these relationships or celebrating them? Given the benefits people can clearly derive from their digital companions, perhaps they are something we could encourage. John MacInnes, the co-founder of MacInnes Scott, which makes hyperreal digital human avatars, describes avatar relationships as perhaps the most important you can have in your life, challenging us to rethink our connection with virtual beings not as one of master and servant but as an ongoing relationship over time that stretches from cradle to grave.

An avatar gives these 'companions' a greater physical presence than any disembodied AI can, providing some much-needed human characteristics that make it easier for people to connect with them: humanising technology, if you will. A key ingredient that is often missing is face-to-face contact. In our traditional interactions with other humans, this has been incredibly important, enhancing our ability to communicate with others and fostering the kind of intimacy that people need. This was illustrated perfectly by a news item during the pandemic about a healthcare worker at Scripps Mercy Hospital in San Diego. Like so many others working in a hospital or healthcare setting, respiratory therapist Robertino Rodriguez had to carry out his duties garbed in a whole-body hazmat suit and a protective face covering. Realising that he could no longer reassure and comfort patients with a simple smile, he took an image of himself without protective gear and used it to create a laminated badge, which he pinned to his plastic overalls. As he made clear in an Instagram post, he needed to establish some relationship with his patients and he realised that there was nothing more powerful than the connection we have when we see another human face.

Neuroscientist and psychologist Lisa Feldman Barrett explains that when you talk to another person, your brain is constantly processing their physical cues so that you can not only respond to what they are saying and doing, you can also predict how the interaction will play out. The more feedback you get – from gestures, facial expressions, the tilt of a head, the look of surprise, a laugh – the more your brain will tweak its model of how to behave. But if you are forced to interact without direct access to these visceral emotional cues, it is much harder to communicate effectively and establish a meaningful relationship with the other person. And this is something that technology firms have also realised when it comes to creating artificial beings that people can connect with.

We're used to our virtual assistants existing as disembodied voices, such as Alexa or Siri, but at the Consumer Electronics Show held in Las Vegas in 2020, Samsung unveiled Neon – a life-like, life-size assistant complete with a computer-generated digital body so that it resembles an artificial human. Each version of Neon can be customised to have its own appearance, its own expressions and its own personality. The company has recognised that face-to-face contact is essential to forming a real connection. As Bob Lian, Director of Strategy for Samsung's STAR Labs, acknowledged, 'human to human interaction is always the best form of interaction'. Thus giving the assistant a real face allows more intuitive and personable communication. And thanks to the convergence of technologies such as machine learning, image recognition, and natural language processing, these assistants can become beings with characters all their own, and respond to humans and their needs in real time so that, as Lian claims, 'you can't tell the difference between them and an actual human'.

This development has implications for a number of industries. Advertising and marketing, for example, where virtual beings are a useful tool but only if they are believable enough to encourage people to buy the product. A simple example can be seen at Rosebud AI, a modelling agency that creates thousands of synthetic beings for use in advertising. They claim to offer 'the most diverse stock photos ever', with some 25,000 models to browse and choose from, all created using AI. The idea is that companies can create exactly the marketing look they want; for example, in a car advert the driver could be programmed as a sixty-five-year-old Caucasian woman for one campaign and a twenty-five-year-old Chinese man in another. The company can use whatever defining characteristics it thinks will appeal to a specific target audience without having to change any other aspect of the advert. In order to make their models feel even more human, Rosebud has created personalities for them and given them a backstory to bring these synthetic personas to life and make them more relatable.

In 2019 virtual reality pioneer Edward Saatchi relaunched his company Fable – moving away from the world of virtual reality film-making to the creation of 'virtual beings', which he defines as 'characters that you know aren't real but with whom you can build a two-way emotional relationship'. He has predicted that one day we'll subscribe to such characters, incorporating them into our daily lives: 'we could watch movies with them, play games with them, cook with them, listen to music with them and also follow their lives on Facebook and Instagram and YouTube.' His own company has developed an endearing digital character for children called Lucy that is capable of having and remembering conversations more complex and meaningful than the question-and-answer dialogue with, say, Alexa. She can ask questions such as 'When was the last time you met a good person?' and 'Did you learn anything from them?' The idea is that children can share insights and stories with Lucy, encouraging them to engage more intimately with her to create a more personal relationship.

Some people might find this strange and question whether we really want to create these sorts of artificial relationships. But the reality is that this is already happening. If we look to Japan we can see that intimacy with virtual characters is seen as much more acceptable than in the West. For the last couple of years, when presenting at events, I have shown a video from a company called Gatebox demonstrating their product of the same name, a cylindrical container that projects a holographic character called Hikari Azuma. She's a cute forever-twenty-year-old, with stripey socks and blue hair, who develops over time through day-to-day interaction with her real-world companion. In the video, Azuma poses and gesticulates from behind the Gatebox glass and communicates with a young Japanese man, who is clearly besotted with her. She wakes him up and wishes him good morning, reminds him of the weather forecast and details of his day ahead, sends him encouraging words by text while he's out and urges him to hurry home. Finally, after he gets home after a long day at work, she greets him with a huge smile and the words: 'Missed you, darling!' Generally, whenever I show it to people, they end up gasping – or laughing – mainly out of embarrassment.

Azuma is the virtual embodiment of Gatebox's guiding principle: 'Living with characters'. She is essentially a 'waifu'

– the name given to the kind of Japanese anime characters that fans would love to marry if only they were real. And, in fact, many men have not let reality get in the way. Gatebox claims to have issued 3,700 certificates for 'cross-dimensional' marriages between its characters and their owners and, given that in Japan one in four men never walks down the aisle, this figure may well increase. Some people question how healthy it is to 'marry' or have any kind of relationship with a virtual being rather than a real-life partner, but others view it as an extension to the concept of morphological freedom (which we'll look at in more detail in Chapter 6). According to one Japanese man, who married his digital partner in 2018, it's just another form of diversity; just as being attracted to someone of the same sex has become acceptable over time, so too will this sort of cross-dimensional relationship.

Japanese cultural critic Hiroki Azuma can give us some insight into the mindset of those who seem to prefer fiction over reality. In 2009, he wrote a book delving into the strange world of *otaku* – the name given to obsessive fans and collectors of manga (comic books) and anime (animated films). When they first emerged in the 1970s, *otaku* were considered to be part of an anti-social subculture, but they are now found all around the globe. In his book *Otaku: Japan's Database Animals* Azuma explains that: 'The *otaku* choose fiction over social reality not because they cannot distinguish between them but rather as a result of having considered which is the more effective for their human relations . . . They choose fiction because it is more effective for smoothing out the process of communication between friends ... and to that extent it is they who may be said to be socially engaged and realistic in Japan today.' In other words, they see the real world as dysfunctional and so construct their own alternative instead, based on their own values and standards.

The 'database' referred to in the title of Azuma's book is an actual database that reflects the obsessive focus that the *otaku* have for these fictional worlds. What's interesting is that the database does not concern itself with the stories that take place in these worlds; instead it concentrates on the various characters and settings that appear in them. *Otaku* call the affection they feel for these characters *chara-moe*, which can be broken down into recognisable defining characteristics called *moe*. For example, the *moe* for character Di Gi Charat include 'hair sticking up like an antenna', 'a tail', 'cat ears' and 'big loose socks' – which along with other characteristics apparently combine to make her cute and huggable. And it is to log, register and store a record of all these elements that the *otaku* database exists – representing an incredibly detailed matrix that documents all the relationships between these features, other characters, quotes, parodies and any other influences. To the *otaku*, this database is like a computer code or a vast collection of recipes and ingredients. Its purpose is not so much to act as a record of who the characters are; it is to give the *otaku* the ability to create their own form of fan fiction. They can lose themselves in rearranging, remixing and reimagining all of the various elements of their beloved fictional universe, constantly expanding and redefining the relationship that they have with the characters over time.

We might already be seeing this trend having an effect on other areas of popular culture. When the electro-synth-pop music artist Grimes released a music video for her song 'You'll Miss Me When I'm Not Around' it came with an unexpected bonus of all the raw files available to download – lyrics, artwork, fonts, video footage and songs stems including the individual guitar, synths and vocals. In an accompanying tweet she said, 'we thought if people are bored and wanna learn new things, we could release the raw components of one of these for anyone who wants to try making stuff using our footage.' It's hard to think of another example of an artist giving up control over all these elements of their work to allow other people to interpret them in their own way. But Grimes is heavily influenced by Japanese anime and manga, so it's possible she may have thought of her musical assets as 'moe-elements' that can be shared, re-worked and re-mixed by her fan communities.

This has the potential to revolutionise the entertainment industry as a whole, allowing the audience to play a more active role in content creation. In the realm of film and television it could enable us to interact with our favourite characters, in a similar way to the *otaku*, and form a relationship of sorts with them. According to Matt Harney of American technology publishing website Hacker Noon, a new form of entertainment 'will be created in your house and broadcast out of it. The celebrity will supply the avatar and then each fan will enjoy a unique experience, depending on their environment and individual reactions. Because of this everyone will be constantly streaming experiences with their favourite characters in case something unique happens.' In this vision of the future, entertainment companies might give birth to a virtual character, but it is you, the audience member, that has the power to direct who that character becomes and the nature of its relationship with you. Linear narratives will give way to something more spontaneous and immediate; character construction will be of far more interest than storyline, as we find ourselves immersed in avatar environments that are not just engaging but totally absorbing.

Perhaps unsurprisingly, Japan has also given rise to an early prototype of this kind of immersion into character through the holographic pop star Hatsune Miku. Having already notched up over 100 million hits on YouTube, Miku owes her success to the fact that fans can generate their own music for her, which she then brings to life and sings for you at one of her 'live' YouTube concerts. I imagine it is only a matter of time before you can arrange for Miku to turn up at a friend's party to sing them a song of your own making. Perhaps these simulated characters will one day deliver a much more personalised form of entertainment than traditional performers ever could.

Customising virtual characters might make them seem more relatable to us, but as we ultimately control them, in a way we would simply be creating an elaborate dialogue with ourselves. That's not to say there isn't value in those interactions. In fact, such an idea lies at the heart of the development of a digital therapy app called ConVRself (originally known as Freud-Me), which allows you to become your own counsellor in a virtual reality setting. You start by choosing a 'self' avatar, then choose a 'counsellor' avatar. Next, embodied as yourself, you explain your problem to the counsellor. Then, switching perspectives to become the counsellor, you listen to yourself explaining your problem. The idea is that the app allows you to consider a problem as if you are an outsider listening to a patient or friend, and it invites you to offer a solution as the counsellor, before listening to it as yourself. As the ConVRself team suggests, you can keep repeating this process as long as you like, and it is claimed to help tackle mental illness, overcome mental blocks and achieve breakthroughs in self-counselling.

At a time of mental-health crisis and ever-tightening public purse strings, perhaps this sort of therapeutic experience could be a useful self-help solution. While we certainly shouldn't overlook the value of an objectively minded, trained professional, for those who can't access such help a VR alternative could prove valuable. Likewise, although I believe we should surround ourselves with a variety of friends with different perspectives, experiences and advice, for those of us who cannot do so, for whatever reason, having an artificial companion to talk to could be the answer. We should still be careful not to do away with too many of our human relationships, as messy and complex as those might be. As Alain de Botton wrote in *The Consolations of Philosophy*, 'We don't exist unless there is someone who can see us existing . . . to be surrounded by friends is constantly to have our identity confirmed.' But I see no reason why some of those friends shouldn't be virtual.

I think the real danger with technology influencing our interactions in the twenty-first century is that it might lead us down a path where we end up as both our 'self' and our 'others' too. It seems to me that many of the human-to-machine relationships that are being developed by tech entrepreneurs could become little more than a series of complex communications with ourselves. While they may start off as interactions between two independent entities, over time the more we are able to customise and control the AI, the more we will gradually turn them into a being with our own preferences, outlooks and behaviours. As Joanne Parsky said to me, 'What we're really doing is reflecting ourselves.' So while we might embrace the development of social robots, friendly avatars and cross-dimensional life partners, what we must guard against is turning these virtual beings into mere mirrors in which we do no more than admire or criticise our 'self', leaving no room for the external influences and experiences that help to shape our identities throughout our lives.

5
REPLACING
YOU

One of the most pored-over definitions of what we mean by 'person' or 'personal identity' must surely be that of seventeenth-century philosopher John Locke. In his magisterial work *An Essay Concerning Human Understanding*, he suggested that the term 'person' be defined as: 'a thinking intelligent being, that has reason and reflection, and can consider itself as itself'. He goes on to suggest a 'thought experiment', asking his readers to consider the case of 'the prince and the cobbler'. What, he asks, would happen if the soul of the prince left the prince's body and entered the body of the cobbler, taking with it all of its 'princely thoughts' and pushing out the cobbler's soul? Locke concludes it is the prince who survives this transformation. A modern version of this scenario is to imagine two men are in an accident, let's call them Peter and Paul. Peter's body has been so damaged it has completely shut down but his brain remains intact. Paul's brain is dead but his body is, miraculously, unscathed. Doctors manage to put together Peter's brain and Paul's body. Which person could be said to have survived this transaction? Most of us believe that it is Peter as his thoughts, memories and reason continue to exist; they are simply encased in a new physical frame.

Of course, our physical bodies do contribute to our sense of identity (which we'll explore more in the next chapter), but most of us tend to agree with Locke that our minds play the truly essential role. What are we to make, then, of the emergence of technology in the twenty-first century that can enhance our minds and our intelligence, from artificial intelligence assistants to devices that can interact directly with our brain? If it is our minds that form the core of our personal identity, what impact could this kind of mind-altering technology have on who we are?

Take, for example, the world of work. Once upon a time, the job we did formed a fundamental part of our identity, forged by the physical tasks that we carried out and the labour skills for which we gained a reputation. It was so fundamental, in fact, that we literally took to identifying ourselves by our occupation. Works like Charles Wareing Bardsley's *Dictionary of English and Welsh Surnames* (1901) contain a whole category of such names. The one that caught my attention first was Napier or 'keeper of the table linen' (one of the below-stairs roles required in a traditional aristocratic household): it was my great grandmother's maiden name and, as is Scottish tradition, it became her daughter's middle name and has been passed down through the family so that it is now one of my own forenames. The dictionary gives plenty of other examples too: Butcher, Butler, Carver, Chamberlain, Cooper, Ewer, Falconer, Farmer, Hunter, Smith, Spencer, Woodward, the list goes on.

Today, while all of these names remain in circulation, many of their associated occupations have become obsolete – evidence of how much social and technological change we have seen over the past few centuries. The rate of change has become so much faster that new occupations emerge and then disappear in almost no time at all. For example, it's not so long since there existed the job role of 'typist'. Pools of women would sit desk by desk, bashing away on metal typewriters to turn out paper copies of dictated memos to be signed and then mailed. Today, while we no longer have dedicated typists, we still have typing. I am typing this manuscript right now but much of what I am doing is carried out by a word processing package that sits on my computer. Advances in both hardware and software mean that we no longer require the hard-won skill and expertise of the old-fashioned typist. In fact, there may soon be no need for any of us to know how to type: it's already possible to dictate your text and have it appear directly on your screen – or someone else's for that matter.

A recent McKinsey report predicted that between 2016 and 2030 the demand for other kinds of physical skills – such as operating a vehicle or packaging up products – 'will fall by 11 per cent overall in the United States and by 16 per cent overall in Europe'. Technology is therefore having an impact on the attributes we choose to develop, reducing the need for physical skills such as strength and manual dexterity, and the cognitive skills required for repetitive roles such as data-inputting or processing. Although people therefore tend to fear that robots are stealing our jobs, the truth is what is happening is not so much a displacement of jobs but a displacement of tasks. The talents that will become ever more in demand are the ones that automation cannot yet replace: higher cognitive skills such as creativity, originality, critical thinking, opinion-forming, ethics and leadership. These are the attributes that draw an identifying line between human and artificial intelligence, and so are the ones each of us might want to hone and protect.

But the truth is that this line is becoming even more blurred all the time, thanks to the ongoing development of artificial intelligence designed to enhance the very same attributes that we would like to place on the human side of the line. In July 2020, Open AI, the artificial intelligence research facility founded by Elon Musk, Peter Thiel, Greg Brockman and others, released an application processing interface (API) allowing users to access a new machine-learning language model called Generative Pre-trained Transformer 3 (GPT-3). Trained on vast amounts of text data from across the internet, GPT-3 is designed to perform a task called 'next word prediction' – a task that could see it perform much more complex roles than you might expect. 'Predicting the next word in a sentence may not sound very useful in a professional context,' says Josh Muncke, whose company Faculty offers consultancy on artificial intelligence to clients including the UK government, 'but what Open AI and others have shown is how powerful this framework can be. It generalises beyond the typical definition of a sentence, and is able to answer difficult physics questions, turn legal-ese into standard English or perform marketing content creation.'

Feed some legal text into GPT-3 and it can auto-complete a legal contract. Prompt it with a few ideas and it can turn out the script for a play. One Silicon Valley investor fed in some starter information on how to run effective board meetings and GPT-3 wrote up a three-step process on how to recruit board members all by itself. 'The iPhone put the world's knowledge into your pocket,' says the investor, 'but GPT-3 provides 10,000 PhDs that are willing to converse on those topics.'

As this kind of high-powered AI assistance becomes more prevalent, it could have a huge effect on the way we work. To begin with, this may mean breaking down our current job roles into AI-assisted micro tasks so that the AI can apply its machine-learning capabilities to train itself in each one. Technology investor Balaji Srinivasan believes that much in the same way that researchers have trained machines to expertly recognise text, so too will professionals such as doctors, lawyers and photographers be able to pass on their own expert skills. In the medical world, for example, he says: 'Taking skilled information, digitising it and pulling it in will eventually allow the doctor to just focus on the really hard cases while auto-diagnosis can take care of other things.'

Josh Muncke says that he can see the potential for what would be a huge step forward in the sophistication of human-to-machine interaction: 'You could imagine fields that involve the synthesis and generation of large amounts of data such as consulting, auditing, conveyancing and customer service being massively upturned by the ability of algorithms to produce 85-per-cent-ready versions of written documents, contracts and other correspondence.' And as subsequent versions of GPT-type models become able to deal with far more complex and contextual requests, perhaps we will reach a time where a manager can simply give a relatively complex instruction – such as 'update the proposal to include a new section on our data protection procedures' – and it would be done. This could also see the emergence of a whole new sector of jobs enabled by and specialised around AI technologies and their application. As Josh Muncke explained: 'We may find GPT analysts and GPT specialists who exist purely to help organisations tune and optimise their in-house versions of these models across a variety of use cases. [Or] we may see organisations or individuals that specialise in producing content – such as marketing – specifically targeted for consumption by these algorithms.'

So, what might this mean for how we think of ourselves and our own potential? Renowned futurist Dr Ian Pearson says that this technology could soon allow us to 'concentrate on the bits [we] want to do and leave the rest to the machine. By starting with hobbies, and bringing them up to professional standards by adding AI capability, we will enable the rise of the polymath. Many people will become highly competent across a range of skills.' He believes that this will have a knock-on effect on the way we work too: '[People] may still have a day job, but also operate on a number of other platforms too. The consequences of this will be that the economy will develop, and so will society. People will start more businesses, business turbulence will increase, and poor-quality businesses will be wiped out.'

Of course, there's nothing to say that these AI assistants would just be geared towards work; perhaps we'll come to use them to maximise our skills, intelligence and efficiency across every aspect of our lives. We may find we need different AI services to match the different ways we wish to present ourselves in different circumstances. For example, you might want an AI assistant that tonally feels like 'you' at work – one that acts and sounds like you at your most expert and professional – and then another that brings out your more relaxed and sociable side for your personal time. And why stop there? Maybe you could have yet another AI assistant for something specific like public speaking to help you come across in as visionary, proactive and communicative a way as possible. Perhaps you might get all three AI services by signing up to a package from the same provider, or maybe you'll find that different tech brands offer a different service, each echoing their own brand personality: an Apple service that emphasises your creative qualities; Amazon that encourages your efficiency and productivity; and Facebook that communicates in ways that never fail to capture the most sociable you. Perhaps this is what Sergey Brin, co-founder of Google, meant when he reportedly said: 'We want Google to be the third half of your brain.'

Some argue that this is a natural extension of the way that we already project different personas in our professional and social lives – a trend that has been accelerated by the advent of social media and its offer of the chance to participate under a pseudonymous cloak of privacy. Balaji Srinivasan explains: 'In the pseudonymous economy, people earn under one name, they speak under another name and their real name is yet a third name . . . Hundreds of millions of people are effectively pseudonymous all day on Reddit and they just use whatever name suits them. They build reputations under their name and they can switch to other names and I think we're going to see a huge version of pseudonymous identities not just for communication which we have already had in the West for years, but for earning.'

I think he's onto something important here. As more and more of us become concerned that our own views might not be in tune with those in authority including our bosses, clients and customers, I think we will become more reticent to express our views under our real names. A pseudonymous identity might therefore come in very handy for expressing viewpoints without affecting the identity we use for work – or even for work with one client rather than another. This vision of a multiple-identity identity might seem weirdly futuristic but it's not so different to the way we have always thought about these things. After all, the word 'person' is derived from the Latin 'persona', which literally means 'mask' – the kind of mask that an actor in Roman times would use in a play. Perhaps we all wear many masks and play many different characters, and the idea of using pseudonymous AI-enabled identities is just a more overt way of organising that.

If we do begin to harness these external artificial intelligences, it will open up a world of possibilities for what we can achieve and how our identities might be shaped by such assistance. But it will raise fundamental questions about the nature of identity itself. For example, where should we draw the boundary between human and machine intelligence? And could a machine ever reach the point where it could be considered – as Locke put it – 'a thinking intelligent being, that has reason and reflection, and can consider itself as itself'? In short, could a machine ever be considered equivalent to a person? Would such a machine be entitled to the same rights as us? Or would we need a new definition of identity that makes us distinctive from these other 'beings'? These are questions that are already being asked by academics and other experts in this field. For example, in his paper 'Legal Personhood for Artificial Intelligence: Citizenship as the Exception to the Rule' bioethicist Tyler L. Jaynes argues that a set of legal protections would have to be given to all non-biological intelligence systems 'in so far as they possess the hardware and software to develop code that surpasses the perceived scope of a human author's initial intent'. Jaynes puts forward three possible legal protections that all non-biological intelligences could be granted in the future:

The right to self-expression, meaning that their observations and opinions would be their own

The right to life where life means containment on electronic systems and is not limited by age

The right to own their necessary components and any other non-biological components they can acquire

There are further rights listed by Jaynes that follow on from these three basic rights, the most interesting of which is the proposal that all non-biological intelligences would have the right to be recognised as a person before the law. This would include – and here the author gives away his US origins – the right to an attorney, to not be a witness against itself in a court of law, and an indictment before a grand jury. Could we one day find ourselves suing an artificial intelligence? Or getting countersued in return?

If you think this all sounds absurd, you should talk to David Gunkel, author of *The Machine Question: Critical Perspectives on AI, Robots and Ethics, and Robot Rights*. Convinced that we must tackle these difficult questions rather than deny that they will ever require our contemplation, Gunkel believes that the idea of AI being given equivalence to a person is perfectly possible – at least in one sense. He argues that 'a person' is merely a recognised legal status, determining someone or something as a legal entity in its own right rather than simply someone else's property. 'In New Zealand they can extend the legal personality to a river, does that mean the river has consciousness? No. It means the river is protected under the same set of rights by which another legal person, whether it's a natural person like you or I, or whether it's an artificial person like a corporation, would also enjoy . . . The same has happened with animals. In India, dolphins are persons.'

I asked David what it would take for an AI to have an identity, not just a legal personhood. His answer was that it depends what we mean by identity, which he believes is more open to interpretation than we might think:

This question is culturally specific and differs across the globe. In the Western European traditions, identity is considered to be the metaphysical property of an individual who persists in his/her/its essential being despite alterations in what the Scholastics called 'accidents.' The famous version of this is Descartes 'cogito ergo sum,' the fact that the identity of being a thinking thing persists through changes in the accidental properties of the alterable body. This is not true in other traditions, like

Ubuntu in southern Africa. In these traditions, personal identity is not some essential metaphysical property you possess or are born with. It is something that develops out of and in response to the community. So 'identity' in these traditions is more of a social and relational concept instead of a metaphysical property. All of this to say that the identity of an AI may be (as it is for us) a matter of cultural location and specificity.

Of course, there are also many scholars who believe that even the most advanced technology that we can develop will never be anything more than a tool; it will never be ascribed the rights or responsibilities of a human person. In any legal dispute involving AI, the responsible parties would be whoever programmed, developed and deployed the AI in the first place. Personally, I find myself falling somewhere between these two camps. I'm persuaded by what Tony J. Prescott, Professor of Cognitive Robotics at the University of Sheffield, calls 'liminality': the idea that we need to define a new kind of being – neither solely mechanical nor quite the same as a biological organism. Even this approach leaves all sorts of questions up in the air: if these systems were to deserve an identity of their own, for example, what impact would that have on how we think about our own identity? And if we no longer thought of these 'beings' as machines, could we continue to treat them in the slave-like way we have always done? Surely we would have to rethink how we lived alongside them. Would they have their own set of identifying credentials? Citizenship? A birth certificate? How would we react if robots took to the streets to demand more rights with the cry of #RobotLivesMatter?

The age of liminality – of beings that are literally on the threshold between man and machine – is in some ways already upon us; after all, people have been walking around with pacemakers for decades. When it comes to having devices like this implanted in our bodies, we do not seem to discern any effect on our sense of identity, but would the same be true if those devices were designed to interact directly with our brains? Scientists have already developed assistive technology such as bionic eyes that transmit impulses along the optic nerve to the brain to partially replace the vision of those who have lost their sight, and AI-enhanced hearing devices that can selectively filter out unwanted noise in order to focus on a particular sound source. How far do technologies like these need to replace or enhance our natural abilities before they affect who we are?

One of the hotly anticipated brain-machine interfaces on the horizon is Elon Musk's Neuralink, which would see a coin-sized implant called a 'Link' fitted into a specially drilled hole in your skull and then covered over by your scalp to make it invisible. On its underside the Link would carry more than a thousand tiny electrodes arranged into threads, which would be laced during the surgery into the surface of your brain. The idea is that these electrodes would be capable of reading neurological impulses and sending signals to the brain, as well as wirelessly linking those impulses to external devices. If successful – and there are many technological, ethical and legal hurdles to overcome – this could allow amputees to control the movement of an artificial limb with their mind, and in the longer term allow us thought-control over external interfaces, so that we could carry out thought-based web searches or messaging as a result.

Musk himself has much greater ambitions, believing that the same kind of technology will eventually allow these threads to penetrate further into the brain and access areas relating to memory formation. He has talked of wanting to stream music to our brains and giving us the ability to save and replay memories. Of course, none of this right now involves consciousness, these are merely smart computers that might be able to simulate some of the connections that we make in the brain. But if every one of the billions of neurons in your brain could one day be connected, it raises the question of whether a device like the Neuralink might allow you to back up your memories and ultimately even download them into a new body or a robot body, an idea we will return to in Chapter 7.

This kind of extraordinary technology has also long been of interest to the military. For over a decade now, for example, the US Defense Advanced Research Projects Agency (DARPA) has funded a project to develop an artificial mind-controlled prosthetic arm, with patients in recent clinical trials describing it as 'like having a hand again'. So far these kinds of projects have involved surgery to implant electrodes, which has limited their use to military or civilian volunteers who have a clinical need. But in 2019 DARPA announced that it was giving grants (rumoured to be around $18 million each) to six teams across industry and academia to develop non-surgical forms of brain-machine interface technology that could be of benefit to a wider population. This new initiative, the Next-Generation Nonsurgical Neurotechnology program (N^3 for short) is more ambitious than anything that has gone before. It is hoped that it will eventually enable new approaches to the management of neurological illnesses, but the end goal for DARPA is to create super-soldiers equipped for a new form of telepathic warfare – able, for example, to control F-35 fighter jets with their minds alone. Futurist Ian Pearson explains the way that such technology might work:

> If you imagine you are trying to enhance the intelligence of a human being, the way you would do it would be to give them some extra thinking space in the IT and to link that to their brain. The human would then see whatever the problem was and be able to process it at high speed on the computer just by thinking about it. Their mind, or part of their mind, would essentially be running on that piece of hardware. In order to do that . . . you have to have the hardware running equivalent to the analogue neural network architecture . . . because it has to run the same processes. And if it's going to be conscious then it needs to be very similar to the way your brain works in

129

a lot of different ways. So once you have that link you would feel like it was your brain but bigger.

For those who think this kind of technology may prove impossible, do not underestimate the surprising plasticity of the human brain. In his books *The Brain* and *Livewired*, Dr David Eagleman explains how the brain is much more influenced by its environment and the information that it gets than we once thought; it is constantly working out how to use this external data to adapt and learn. He argues that we should therefore view the brain as an extremely fluid system, and that if we can plug in new kinds of data streams our brains will just work out how to use them. In fact, Eagleman has built a company called Neosensory to explore this potential, and has already released the Buzz wristband which allows deaf people to 'feel sound on their skin'. Proof positive of the brain's ability to make sense of new forms of data, the wristband contains vibratory motors that convert audio streams into sensory patterns on the skin, allowing deaf people to 'hear' sounds and distinguish between them by recognising the particular rhythm of the vibration. Neosensory has now built on this system to allow one person's brain to become aware of another person's physiology, such as their heart rate or galvanic skin response. It works by taking the data from your smartwatch and connecting it not just to your Buzz wristband but to another person's wristband too. 'Imagine you and your husband are somewhere else in different locations,' says

Dr Eagleman, 'you could feel your husband's heartbeat, you'll know if he's feeling stressed and you might want to call and check in on him.' In the past we may have wondered what it was like to be in someone else's body. It's early days for this technology but we are now starting to get a glimpse of just that, as our intelligence begins to make sense not just of our own thoughts and physiology but that of others too.

The exciting thing is that we do not yet know the limits of the brain's plasticity. If technology such as the Buzz wristband is already possible, why couldn't we train our brain to not only receive information but send out information to wirelessly control a machine from across the room? Perhaps – in the same way that DARPA hopes to link soldiers to fighter jets – we too might find ourselves able to thought-control any kind of everyday machine from a vacuum cleaner or a smartphone to a more sophisticated workplace robot or even robots in outer space.

Then things get even smarter. Once we are able to plug our brains directly into specific devices, imagine plugging ourselves into some kind of shared server onto which we could extend our mind, offload tasks, store memories, increase our mental capacity, and even connect with the brains of other people connected to the same server. When I asked Ian Pearson to explain how that could work, he responded:

Let's suppose we get to the point in 2040 or 2045 where direct brain links are starting to appear. They can't upload your entire mind but they can act as an extension of your memory, so an extension of your intelligence and some of your mind can then run on the cloud. [If] your mind is running on the server farm, and my mind is running on the same . . . farm, there's nothing in principle to stop you and I from exploring the same concepts at exactly the same time because our brains start to overlap at that point; your brain includes this particular chip or this particular algorithm and so does mine. And because we share the same thinking space we could then share ideas and develop ideas together without directly talking to each other.

This raises so many extraordinary questions for us to grapple with. Does this mean we could effectively watch the thoughts of every other person whose mind was relying on the same server farm as us? What about when we're sleeping – would our extended minds still be working away and does that mean someone else might be able to access our dreams? And if that is the case, does that mean our dreams will no longer belong to us, but will become as public as any other media is today? Would we be looking at a new era of social media, in which news feeds are replaced by 'dream feeds'? If that ever becomes the case, I'd think we'd all get very familiar very quickly with the privacy settings!

It would also be transformative for many industries – especially in creative fields such as advertising, marketing and design. Those industries love to talk about the 'creative idea' and the collaboration it takes to originate a big idea and then refine it through a series of drafts and redrafts passed back and forth between designers, writers, account managers, executives and clients until it's perfected. If we ever reach the era of the shared mind, would we be able to get rid of all these interim stages? Perhaps all of the people involved wouldn't have to depict their ideas and desired corrections in order to get input from someone else, they could just all think the idea through at once in the mega mind. It would be, as Ian Pearson says, 'like using Google Docs where more than one person can edit the same document at the same time . . . [it would be] Google Mind Docs'. And it could be two people interacting, or it could be two million – and then what you would have is a hive mind.

This notion of the hive mind is often associated with a movement called transhumanism, whose supporters are united by the belief that humans can and should transcend their natural biological 'restrictions' using science and technology. We'll return to the topic of transhumanism and biological enhancement in Chapter 6, but its philosophy has interesting ramifications for psychological enhancement too. Some transhumanists believe that if we can plug our individual minds into the kind of shared technology described above – and so unhook ourselves from the confines of our biology – we may one day be able to 'live' in a bigger, more intelligent, integrated consciousness, a 'collective consciousness' if you will.

Zoltan Istvan, one of the leading figures of the transhumanist movement in the USA, thinks that such a fundamental change to the way we think about our personal and collective identity is not so far over the horizon:

> Twenty or thirty years into the future, with an implant in your brain, or forty or fifty years [when] you can connect with the cloud, you will be able to connect with your spouse and your children directly . . . it'll be instantaneous, like we get tweets on our phone or direct messages all the time . . . We're going to be much more integrated with one another, we might not even have an identity. One hundred years in the future I would be very surprised if it's just Zoltan. I think it will be Zoltan and his tribe: my wife, my kids, their children, my grandmother and father.

To me this would be anathema because I cannot accept the idea of not having autonomy over who I am; any sense of individuality, any sense of identity, would have all but disappeared. It's bad enough *watching* the mob surge and groupthink on shared social platforms like Twitter, but imagine what it would feel like to have your mind fully caught up in it. 'Who knows what it will be like?' says Zoltan but he thinks we need to be prepared for a future in which – maybe just sixty years from now – privacy has been consigned to the past. He believes we need to ask ourselves how close we want to be to our loved ones. Do we want to know everything about our partners or our children? 'I have a nine-year-old daughter,' he says, 'and she's just starting to go out and walk herself to school. Does she want me in her mind? These are very challenging questions, and I think with a lot of parents, we're going to have to decide how much we want to be in each other's space. I don't mean physically, I mean mentally, it's a very different space to be in someone's mind.'

I can't help wondering about the practicalities of all of this. What would happen, for example, if you wanted to get divorced (I'm guessing it would be pretty impossible to hide an affair)? Would you be able to decide who else gets to be part of your hive mind, or would you need the agreements of the others in your mind tribe? Who would get custody of which memories? Or would the memories all be copied so that everyone leaves with the same mental photograph album of their family history? Could anyone be erased from that? And if you are allowed to leave, where would your mind go: would it return to the group it was originally linked to, say your birth family? Moreover, how would you maintain two or more hive minds, one for working with colleagues and one with your family? Presumably the two would need to remain discrete. They would certainly need to be made extremely secure, as hackers who gained access to your personal hive mind group could potentially gain access to the thoughts of your children.

As Zoltan says, these are questions that it's worth us all thinking about now. Perhaps the technology we're talking about is decades away, but we would be foolish to under-estimate the pace and potential of change, and its possible implications. No one would deny that technology has already altered the relationship between our identity and our working lives – a trend that stretches back for centuries. Today we face far more fundamental challenges to the way we think about who we are – whether it's the development of devices that can expand the capacity of our brains, robots that we can control with our minds, and AI that could justifiably claim an iden-tity of its own. Perhaps even more fundamentally, we need to ask how we can maintain our individuality in a future world in which we become intrinsically connected to each other, where one brain might control several bodies or one body might have access to multiple brains. In such a scenario, per-haps the very concept of personal identity could not survive.

6
ENHANCING
YOU

In Bernard Williams' 1970 paper 'The Self and the Future', he sets out a thought experiment: imagine you are a prisoner and your captor tells you that you are going to be tortured tomorrow. But he also tells you that you will not directly experience or remember it as your mind will be replaced with someone else's beforehand. Despite that caveat, most of us would still be scared for our physical self. As Williams put it, 'No amount of change in my character or beliefs would seem to affect substantially the nastiness of tortures applied to me; correspondingly, no degree of predicted change in character and beliefs can unseat the fear of torture which, together with those changes, is predicted for me.' That being the case, perhaps our identity is not associated only with our mind, as suggested by Locke in Chapter 5.

Indeed, we are reminded of the physicality of identity by the touching stories that surround successful transplant operations. Sadly, many of the healthy organs required for such operations come from young people who have been killed in some kind of accident, and it is common for the donor's relatives to say that they feel their loved one is still present in the world thanks to the transplant. As one bereaved parent put it, after using a stethoscope to listen to their little boy's heart beating in another child's chest: 'Part of him is still going to be living on.' And that 'part' of their child's identity is contained in the tangible, physical organ that remains very much alive.

There is nothing especially surprising about the way we associate our physical make-up with our sense of identity. If asked to differentiate ourselves from others, it is natural to point to our age, gender and height and the colour of our eyes, hair or skin. Indeed, this association is perhaps stronger than ever thanks to one of the most significant scientific breakthroughs in modern times: the publication in 2003 of a fully sequenced human genome. Often described as the body's instruction manual, your genome is the sequence of over three billion genetic variants that give you your biological uniqueness. There is a copy of it in every healthy cell in your body, and to a greater or lesser degree it determines your eye colour, height, propensity to disease and even how long you may be able to stay on this earth. In some cases, analysis of your genome can also be used to detect any predisposition you might have for a specific disease or to predict how well you will respond to a particular medical treatment.

It is, in a sense, a treasure trove of data on what makes you 'you' – and data is what lies behind the drive towards ever more precise and personalised healthcare.

We have already seen a huge increase in the amount of personal biological data that we can capture, store and analyse about our bodies. Today, thanks to the advent of wearable technology, wireless communication and data-processing power, we can keep this kind of information flowing and updating throughout every minute of every day, giving us new insight into our patterns of behaviour, and the way that our decisions and our surrounding environment affect our health over time. Chris Dancy, often referred to as 'the most connected man on earth' due to the amount of wearable tracking technology he uses, believes that this is changing the way that we think about our identity. 'I realised very quickly,' he says, 'that my identity was being captured and catalogued and narrated in so many ways and moments [and] those moments were data rich . . . Wearables are popular not because they are counting steps but because . . . they are showing you who you are.'

Given the trend for data-driven healthcare – and the fact that the cost of sequencing a whole human genome has fallen dramatically from $10 million in 2006 to somewhere around $500 today – it is no surprise that people are being encouraged to have their genetic data assessed to understand their own genealogy, propensity to disease and personal wellbeing. Thanks to services such as ancestry.com and 23andMe, there is a growing market for 'direct-to-consumer' testing to enable anything from tracing a family tree, determining paternity or quantifying your risk of developing a degenerative disease such as Alzheimer's or Parkinson's. As the Nuffield Council on Bioethics noted in its submission to the UK House of Commons Science and Technology Committee in April 2019: 'Developments in commercial genomics can provide tools to enable people to take more responsibility for their health', adding that it has the potential to 'allow early intervention, give more personal control of one's health, save public health-care resources . . . and alert relatives to important genetic conditions or predispositions.'

To find out more, I decided to take a genetic test, which incorporated hormonal and nutritional analysis, with Pippa Campbell, a UK-based nutritionist and practitioner in functional medicine whose popular information-packed posts about health, wellness and nutrition attract many fans on Instagram. She partners with a company called Lifecode Gx to offer DNA testing (via a simple cheek swab) and a support package covering a variety of targeted tests. I found the results of my analysis fascinating and feel far more informed about how my particular body works – especially in relation to my liver and my adrenals. I learned, for example, that although my oestrogen levels are dropping, what is being produced is not being eliminated in the right way and that my body is finding it hard to detoxify it, partly because of a particular mutation in one of my genes. Coupled with the discovery that my oestrogen receptors are too sensitive, this means that I now know not to ever undertake hormone replacement therapy; it would simply flood my body with additional oestrogen that I can't detoxify, which would be the very worst thing for me.

Following my analysis and armed with DNA-informed advice on food, nutrition and supplements, I lost nearly two stone. And according to Pippa, there can be advantages when it comes to mental health advice too. 'A lot of psychiatrists are now training with Lifecode Gx,' she says, 'which have some very specific genes that help us understand medications and painkillers and how quickly different people detoxify them. Then, the psychiatrists can use these reports to understand what will work and what dosing to prescribe.'

This is the kind of personalised health analysis I think we all want, and that national governments would do well to offer their citizens. In the UK, the NHS's Long Term Plan does envisage a more personalised care model and such an approach is being piloted in some parts of England, but it is focused on the management of existing illnesses rather than a preventative health programme. In the long run, it is surely more economical to provide individuals with their own ana-lysis, and design personalised nutritional, supplemental and fitness programmes for each, rather than spending billions telling *everyone* not to eat too much, and to get on a bike instead of taking the car.

In 2012 the UK government announced the launch of 'The 100,000 Genomes Project' – an ambitious scheme to sequence the genomes of 100,000 NHS patients, both children and adults, something that no one in the world had yet attempted. By December 2018, it had reached its goal, amassing a huge amount of genomic data, with a particular focus on patients with rare diseases and some common kinds of cancer. The primary aim of the project was not to provide diagnostic benefits to the patients who took part but to provide a research resource that would help scientists harness the power of genomics to the benefit of all future patients. It was an enormous undertaking, but it also involved a good deal of ethical complexity. For example, parents whose child was participating in the project could – for the first time – opt to find out not only whether their child had the genetic variants associated with their specific condition but also whether they had variants relating to their susceptibility to other diseases. This has raised questions about how this kind of data could and should be used – not least because the same technology could pave the way for parents and practitioners to embark on the production of genetically designed offspring.

Joyce Harper, Professor of Reproductive Science at University College London, has worked in the field of in-vitro-fertilisation (IVF) – and specifically pre-implantation genetic testing (PGT) – for about twenty-five years. She explains that when embryos are created in the lab, they are also tested for any genetic disease that the parents could carry so that only embryos free from disease are transferred back to the mother. In some cases, where there are viable embryos of both sexes, a couple may also be given the option to select the sex of their child. Professor Harper thinks that in the future, just as adults can pay commercial firms to have their genome sequenced today, every embryo may have its genome sequenced so that decisions can be made about what characteristics, traits and skills the parents are looking for in their child. Speaking at an Oxford University Union debate, she highlighted some of the issues involved by using the example of two concert pianists who want a child with perfect pitch. 'Perhaps it would be possible in the future to only select the embryos . . . that are carrying absolute pitch,' she said. 'But what if that child . . . doesn't ever want to play piano? What if they want to be a footballer instead?' As she points out, we all know parents who force their own expectations onto a child, but this would take it to another level.

We are presented with a similar scenario in *Hacking Darwin*, a book by Jamie Metzl, who was appointed in 2019 to the World Health Organization expert advisory committee on human genome editing. Metzl asks us to imagine a scene in a fertility clinic of the future, where an IVF patient and doctor are discussing how far the pre-implantation genetic testing of her embryos should be taken. The conversation moves from an everyday discussion about single gene mutation disorders like Huntington's disease or cystic fibrosis to the less ethically straightforward notion of influencing traits such as gender and hair colour. And, based on his belief that we will eventually know much more about the effects of multiple genes working together, Metzl then takes the discussion into even more complex territory such as IQ. Surely, he suggests, any parent would want to give their child the best possible chance in life by giving it advantages such as a high IQ?

According to Metzl, the decisions we would make in such a scenario are likely to be motivated by loss aversion. If you had the chance to increase your child's future chances in all of these ways and more, wouldn't you do it? Wouldn't *not* giving your child these possible advantages at this stage be tantamount to disadvantaging it later in life, because every other parent in a similar position would most likely be opting to enhance their child's traits, talents and life chances at this stage too? Right now, these are hypothetical questions, but the scenario that Metzl presents to us is nonetheless a plausible one. We need to consider what effect these kinds of decision will have on the way we think about identity. Is it desirable or even ethical to design a person in this way, choosing features and characteristics for them that will affect not only who they are but how they think about themselves?

On the one hand, the idea that we could use genetic intervention to write a child's future can be seen as a way of opening up that child's opportunity to live a more fulfilling, rich, long and healthy life. But on the other, it has the potential to close down future possibilities. There may be characteristics that parents today might perceive to be valuable and to fit our present-day norms that in the future their children might come to regard as a disadvantage. As Joyce Harper muses: 'Will parents become like programmers designing their children around the particular characteristics that they desire? What about if we have fashions like the body shape of the Kardashians but then when the children grow up these fashions are no longer in mode. Will the children sue their parents for lack of judgement?'

American futurist Roy Amara came up with what is known as Amara's Law; it states the tendency to overestimate the short-term effects of technology and underestimate the long-term effects. So while there is plenty of water to go under the bridge before individuals are faced with making such decisions – not least because we are only just beginning to understand how our genomes work – we should start considering the longer-term implications now. It is one thing, for example, to try to fix a disease caused by a mutation in a single gene or to alleviate functional problems that might be caused by genetic pathways but, as futurist Dr Ian Pearson explains, the vast majority of human traits are governed by a much more complex genetic picture: 'The problem is that most things in your body aren't just covered by one gene. They're caused by the interactions of a number of genes. That number could be two or it could be twenty, or it could be two hundred, and in most cases we just don't know. It's quite easy to discover that a particular gene is associated with eye colour or hair colour. It is not very easy to know whether that gene might be involved in anything else as well.'

Despite these complexities, scientists have succeeded in developing biological techniques to identify and characterise genomic targets and then make 'edits' to those target cells. More than one technology exists to carry this out but one in particular has stolen the limelight. It is called CRISPR, which stands for 'Clustered Interspaced Short Palindromic Repeats', or more technically CRISPR-Cas9, where 'Cas' stands for 'CRISPR-associated protein'. Pioneered by Jennifer Doudna and Emmanuelle Charpentier (earning them the Nobel Prize in Chemistry in 2020), the CRISPR technique makes use of the Cas9 protein to add and remove genes from the cells of living organisms. And it does so in a way that has proved efficient, accurate and 'easy to use'.

Perhaps too easy. In 2018, Chinese scientist He Jiankui claimed to have used CRISPR-Cas9 to edit the genes of two foetuses in order to give them resistance to HIV. More than that, he had gone on to implant both foetuses, leading to the birth of twin girls – the world's first gene-edited humans. Given that the CRISPR technique is still in the earliest stages of basic research, He Jiankui's decision to use it for reproductive purposes was a gross violation of ethics, regulations and laws, and it provoked outrage from experts around the world. Jiankui has since been fired from his job and sentenced to three years in prison, but perhaps he has given us a glimpse of a possible future – whether it's one we want or not.

Back in the West, technology is changing the way we think about some of the most fundamental aspects of reproduction, sex and gender – both as individuals and as societies. And this in turn is changing the way we think about identity. Take, for example, Martine Rothblatt, the American entrepreneur and top-earning CEO in the biopharmaceutical industry in 2018. She believes that technology is putting paid to what she calls 'the age-old apartheid of sex' – that we will one day stop 'labelling' people as male or female at birth. How so? Because, in her view, 'sex lies at the heart of biology and yet in transcending biology, technology [has given] us an explosion of sexual identities.'

We are living in a world where more and more people, like Rothblatt, are questioning the traditional binary approach to gender labelling. Transgender people, for example, have a gender identity or expression that differs from the sex they were assigned at birth. And medical science has now made it possible for transgender people to transition from one sex to another. But, while people can make a great many changes to their biology, they cannot yet adopt different reproductive abilities. So, while pregnancy is possible for a transgender man who has undergone female-to-male transitioning (as long as there is a functioning uterus, ovaries and vagina), it is not yet possible for a transgender woman following male-to-female transitioning. However, that too might change in the future.

In 2017, a US woman successfully gave birth to a baby after receiving a uterus transplant, and a further fifteen women are known to have had the same surgery since, with eight giving birth, all in Sweden. It is still early days for this form of surgery, and there are considerable potential complications for both mother and baby, but it opens up the possibility that the same treatment might one day be available to implant a uterus and then a baby in a transgender woman. The implications of this would be huge and would undoubtedly spark a great deal of serious discussion and debate at a societal level, but what would it mean for our purposes? How would it affect the way we think about individual identity? We got a taste of the questions we will have to wrestle with back in 2007, when transgender man Thomas Beatie hit the headlines as – in the words of the press – the 'Pregnant Man'. As many asked at the time – after he successfully gave birth to a baby girl – can someone capable of giving birth be considered 'male'? Is Thomas the baby's father or mother or both or neither? Perhaps, as these traditional labels begin to fall short, we will have to define 'motherhood' and 'fatherhood' by something other than the chromosomes and sex organs of the person involved. Some argue that we will talk less about nouns, and more about verbs such as 'mothering' and 'fathering' – actions that can be carried out by all regardless of sex or gender. Or perhaps we'll need some new vocabulary to transcend biology altogether.

Certainly Martine Rothblatt, as a self-declared transhumanist, has talked about technology liberating us from our biology in relation to, though not confined to, our sex and gender. Ideas about transhumanism have been around for about a century. They were first advanced in 1923 by the British geneticist J. B. S. Haldane, who was interested in how the field of genetics could over time be used to alter human characteristics, but evolutionary biologist Julian Huxley is generally considered to be the founder of the field after publishing an influential essay in 1957 on the future of humanity. At that time, of course, there were growing fears that humanity may not have much of a future, thanks to the heightening political and military tensions of the Cold War. And it was in the aftermath of the Cold War that transhumanist theory began to translate into intentional policy.

According to Professor Steve Fuller, one of the world's foremost academic proponents of transhumanism, this tipping point came with the publication in 2002 of a report called 'Converging Technologies for Improving Human Performance' by the US National Science Foundation and Department of Commerce. Examining the potential for the fields of biotechnology, information technology, nanotechnology and cognitive science to improve our mental and physical performance, the report recommended that such endeavours should become a national priority. 'This is the transhuman moment,' says Professor Fuller, pointing also to the influential 'Converging Technologies' report published by the EU in 2006.

Both reports focus on how we can harness technology to enhance the human condition, whether it's helping people to live longer and healthier lives or seeking to make populations immune to the threat of biological warfare. And in Fuller's interpretation, that is what transhumanism is about – a way to enhance our humanity, to prolong our life in a way that ensures we aren't holding ourselves back and squandering our potential. It is founded on the principle that humans are not just one of many species in nature but a privileged kind of species capable of much more resilience and progression than any other. And if we have that capability, why wouldn't we use it?

Given that transhumanism is all about redefining what it means to be human, it obviously has a great deal of relevance to the question of identity. Professor Fuller points in particular to the idea of morphological freedom, which he describes as 'the idea that human beings have the right to *be* in whatever form they want'. Explaining what this means, Professor Fuller suggests that we might think of our physical bodies as just one possible 'platform' for living our lives and that we should consider the possibility that we can 'migrate across platforms' in order to enhance or extend our existence. 'If you want to upload your mind . . . you can do that; if you want to [stay] in your own body you can do that. [It's] a very ontologically tolerant position.'

It certainly seems as if the tide of public opinion is turning that way. A survey carried out by market research and insight company Opinium in 2020 suggested that 63 per cent of people in Western European countries would consider augmenting their bodies with technology, with most believing that it would improve their quality of life. My own point of view is that many of the people born into the world today will by the middle of the century actively identify as transhumanists. One of them, I strongly suspect, will be the baby boy born to technology entrepreneur Elon Musk and musician Grimes in May 2020. Announcing the birth, the proud parents revealed that they were calling the boy X Æ A-12. Admittedly, no one seems to know how to pronounce this – including Musk himself – but at least Grimes explained what each of the symbols means via Twitter:

- X, the unknown variable
- Æ, my elven spelling of AI (love &/or Artificial intelligence)
- A-12 = precursor to SR-17 (our favorite aircraft). No weapons, no defences, just speed. Great in battle, but non-violent.

If ever there has been a transhumanist baby name, then this must surely be it. Naming a newborn using mathematical symbols, fantastical languages and military machinery may seem like a mere gesture, but I think it hints at the way in which innovative people with unusual sensibilities are starting to envisage a different type of future human, a more techno-logical being. It is certainly a first – and perhaps the world isn't ready for it: the baby's name was later changed to X Æ A-Xii when it fell foul of state law in California, which states that only the twenty-six letters of the English alphabet can be used to register a name on a birth certificate. Another case, perhaps, of our governance structures lagging behind and restricting our human potential!

The idea of a more technological human has long been a staple of science fiction, but the truth is that such humans are already living among us. Kevin Warwick, for example, is a long-time cyborg. A professor of cybernetics at Reading University in the UK, in 1998 he became the first person in the world (as far as we know) to have a silicon chip surgically implanted in his body. He saw it as a permanent smart card that would never be lost or stolen because it had become part of his 'self'. And he imagined that it could one day include information about his blood type, national insurance number, qualifications and even criminal convictions or speeding fines – creating a biological version of the kind of digital identity credentials that we discussed in Chapter 1. At the time of the surgery, Warwick was quoted as saying: 'Human intelligence needs to keep up with the intelligence of machines. If we implant silicon chips into humans, humans can stay ahead.' And speaking in December 2019 on the FUTURES podcast hosted by science communicator Luke Robert Mason, he reiterated that he couldn't see a reason why we wouldn't be chipped from birth: 'Just an RFID [radio frequency identification chip] as a passport . . . has benefits. You might need an upgrade from time to time . . . But it'd be good to have it at birth. As a person develops at birth, they develop with that new sense as though it's always been there.' He then went even further, saying that he believed it was human nature to embrace such technology:

[People will] say: 'Well, I have the right to upgrade, as an individual. To upgrade if I want an implant like this. To have some of those abilities.' And humans . . . from when humans [first] appeared . . . have gone on to improve ourselves. We started to fly. We drive cars. We have all sorts of technology, so humans will go for it. Humans will go for the upgrade.

One man who has already opted for the upgrade is Nikolas Badminton, a futurist in Canada, who decided to get a microchip implanted into his arm. Manufactured by a company called Dangerous Things, the chip can be used to connect to key fobs and swipe cards, password readers and other devices through passive RFID. When I spoke to him, Nikolas told me that he doesn't even notice that the chip is there. He said:

The idea of implanting a microchip is more than just the current idea of turning a human into an access control device – keycode, URL, bitcoin cypher, or otherwise. It's the beginning of a journey towards Human 3.0 where we may realise true human biological symbiosis with technology and develop the abilities to upload information, augment our intellect, extend our bodies' physical abilities, or become biological depositories of data.

When it comes to enhancing our bodies, I think it's likely that the initial technology of choice will be smart glasses rather than implanted chips. One of the other questions in the poll conducted by Opinium asked what physical attributes people would most want to enhance, and while 40 per cent of respondents opted for overall physical health, the most popular area for specific improvement was eyesight, at 33 per cent. I think this demand might help to make smart glasses the gateway tech to augmented humans. Unlike a smartphone in your pocket, they would always be on your face, surveying the environment – capable of recording and storing all you see and hear and even uploading clips straight to your social feed.

But while many of us find this kind of technology exciting to talk about, we are less inclined to get into a discussion about who would own the hardware and software involved, or the wealth of associated data. In the same way that the fictional Tyrell Corporation asserted ownership of its bio-enhanced quasi-human 'replicants' in the *Bladerunner* films, it is quite possible that a technologically enhanced human will have very little independence from the corporation that will 'own' and enable his or her enhancements. After all, as consumers we are extremely lackadaisical about reading the terms and conditions for a new smartphone before we sign up to use it. Why should it be any different in the future? Perhaps you would opt to have a shiny new chip implanted without fully realising quite what you were agreeing to. It might include having to pay for routine upgrades to keep your implant functioning properly and allowing the company to make use of your data for commercial purposes. In a case such as this, who would have control of your data and the services that the technology enables – you or the corporation? Is this transhumanist libertarian road really a route to freedom at all?

In July 2020 the World Economic Forum published a report intriguingly entitled 'Shaping the Future of the Internet of Bodies', which sketched out all the different ways in which medical and non-medical enhancements might lead to a brand new set of governance challenges. On the question of how the data associated with these enhancements might be used, the report says: 'Some experts advocate a solidarity approach to health data governance (one focused on societal and community good). This shifts the focus to the shared societal benefits and responsibilities, which motivates people to share data for the collective and individual good. Biobanks are a good example for sharing biological data.' In other words, this suggests that while it may not be deemed mandatory, there will be pressure on individuals from some quarters to make their biological data available to the technology platform or the state system 'for the common good'.

Setting aside these potential future issues of governance, transhumanists have other hurdles to overcome before their ideas become more mainstream. Zoltan Istvan outlined three main challenges that are preventing the transhumanist philosophy from becoming more widely accepted. The first, he said, is religion, which he thinks tells us 'don't touch the body, it's sacred', but he is hopeful that the coronavirus pandemic might help people realise just how life-limiting it is not to prioritise science. The second is a lack of understanding of the technology that already exists today, and how quickly it is developing, which means that people are too quick to dismiss transhumanism as science fiction. He said, for example, that 'people don't realise that . . . [by 2028] there's a very good chance that a quadriplegic in an exoskeleton suit will run dramatically faster than the fastest runner on planet earth.'

The third factor that Zoltan identified is the way that institutional conservatism serves to hold up technological advancement. 'Genetic editing could be beneficial for every single person alive,' he said, 'yet what you hear is that this is something dangerous, we need a moratorium on it, we need years of considering it.' But such delays are harmful in themselves, according to Zoltan, who argues that by holding back from the use of genetic technology, we are causing the unnecessary deaths of 150,000 people every day. It is a thought-provoking claim but critics would argue that such caution is necessary when we are talking about technology that could lead to such a fundamental change in what it means to be human.

One of the key issues for transhumanists to contend with is how the adoption of technology might lead to greater social inequality. In 2015 I attended the World Future Society conference in San Francisco, where I heard a talk by one of the most renowned futurists living today, Peter Schwartz. Talking about precisely this subject, Schwartz predicted that one of the biggest disruptions to take place over the next 100 years would be 'the speciation of the human species', which he defined as the intentional differentiation not just between species but within species to create a greater variety of forms. He said that he anticipated that the human species would, for the first time ever, be able to control its evolution and that the next big evolutionary step would be how technologies such as genetic engineering could be used to increase the diversity of future populations.

According to the website Space.com, this extraordinary vision of our future has since moved closer. In June 2020, it reported on the first summary results of a study conducted by scientists on how to genetically engineer humans capable of surviving a mission to Mars. The study, which concentrated on twin American astronauts Mark and Scott Kelly, was designed to explore the effect of space on our genetics. It did so by comparing Scott's DNA before and after he spent a year on the International Space Station with the DNA of his twin brother Mark, who spent the same year on earth. Now, having uncovered various insights from those experiments, the team of scientists is exploring ways to mitigate the effects of long-term space travel, with one possibility being to combine human DNA with that of another species. Among the candidate species are some aquatic invertebrates called tardigrades – also known as water bears. Tardigrades have the resilience to survive in all sorts of extreme conditions, and the scientists are hoping to pinpoint whether there is something in their genetic make-up that helps them to better protect and repair their DNA. If so, it might be possible to transfer that trait to human cells so that astronauts can withstand – for example – the effects of exposure to radiation. And of course that would not only be immensely helpful for space travel but for other aspects of our lives here on earth. As one of the lead scientists on the project puts it: 'It's not if we evolve, it's when we evolve.'

But how will we evolve and – perhaps more importantly – which of us will get to do so? Once such technological enhancements become available, isn't it likely that they will only be available to the wealthy? This is already true of IVF treatment today, which in the UK – unless you are lucky enough to receive NHS funding – can cost around £7,000 per round and about £5,000 to freeze your eggs. If genomic editing in search of preferred characteristics does become a trend, I imagine we can expect to see a similar pattern in terms of affordability. This then begs the question whether the wealthy will start to think of their prospective children as some kind of consumer choice, something to be custom-ised, with no expense spared. And if you aren't wealthy, well, your children will be left behind, un-enhanced and potentially uncompetitive, struggling to match the biological advantages conferred on their more fortunate peers. In other words, this would be 'survival of the fittest' but where 'fittest' might refer to qualities that are as much technological in origin as they are biological.

Another concern is that in our money-oriented societies every 'upgrade' that people might opt for will in the end be designed to give them higher and higher productivity, until in the end the technologically enhanced human becomes little more than a unit of capitalism, creating and producing and forever upgrading to keep up with their expected output. We saw in Chapter 2 that even a simple immunisation certificate has the potential to drive a wedge between different groups in society, between those who have some kind of technological access to what life offers and those who don't. Imagine how much this kind of inequality would be magnified if there were different versions of humans within a society: some who have the resources to pay for their regular technology upgrades, and some who do not. Suddenly the idea of being 100 per cent human would be seen as a disadvantage – an inferior form of being. Consider too the possibility that technologically enabled humans might one day be able to increase this gap by downloading additional educational abilities. Is trans-humanism for us all, or is it just for a wealthy elite?

Conversely, those who do choose to be technologically enhanced could face a different kind of disadvantage – the loss of privacy and personal freedom. Once you start weaving any kind of technology into your biological make-up, you are running the risk of that technology being hacked. Yuval Noah Harari outlined just such a scenario in an explosive speech at a meeting of the World Economic Forum at Davos in January 2020:

If you know enough biology and have enough computing power and data, you can hack my body and my brain and my life, and you can understand me better than I understand myself. You can know my personality type, my political views, my sexual preferences, my mental weaknesses, my deepest fears and hopes . . . In the past, many governments and tyrants wanted to do it, but . . . soon at least some corporations and governments will be able to systematically hack all the people. We humans should get used to the idea that we are no longer mysterious souls – we are now hackable animals. That's what we are.

Asking his audience to imagine what would happen if this power were to fall into the hands of a 'twenty-first-century Stalin', Harari warned that if we do not grasp the 'new hell' of 'digital dictatorships' quickly enough, we will find ourselves trapped with no way to escape.

It is a warning that we should take seriously. After all, governments already know more about us now than they ever did before – whether through surveillance or through the kind of genetic data that we willingly give up through health-driven initiatives such as the 100,000 Human Genome Project. But when does health management become citizen management? In June 2020 the *New York Times* reported that police in China have been collecting blood samples from men and boys – including those of pre-school age – across the country to build up a genetic database of its 700-million-strong male population. The collection, storage and use of genomic data in this way is illegal according to international law and even China's own criminal law, but the report also alleges that China's Ministry of Public Security began building a forensic genetic database as far back as 2003 and that it has been collecting DNA samples from those unconnected to criminal activity and without their consent since 2013.

Setting aside the question of legality, what could be the purpose of collecting such a database? Perhaps it could be used for the speedy and accurate identification of criminals and so deter such behaviour, but it could also be put to much more sinister uses. Having such a comprehensive database would help enormously, for example, in the manufacture of viruses, bacteria and other bio-weapons to target specific groups. Not only that but if a citizen's DNA were to be linked to their social credit score, the state would effectively have created a biological breeding score. Futurist Dr Ian Pearson outlines a frightening scenario in which 'you might not be allowed to breed unless you've got the particular genes the state wants', which would mean the government could 'decide what your population looks like'.

And this dystopian scenario could apply to any country in the world that is governed by an autocratic regime. If a government wanted to implicitly or explicitly favour or oppress particular groups in society – whether related to race, class, sexuality, gender, religion, immigration status, physical abilities, mental health and even body shape and size – it might soon have the power to do so more effectively than ever before. With a powerful genetic database, rapid advancements in genetic editing, the development of chemicals to target particular gene carriers, the outcome could be a population that expresses only the characteristics and traits that the state deems acceptable. Your identity would no longer be something that you can evolve for yourself over time, nor even anything your community can bestow on you; it would be whatever successfully survived the environmental conditions set by a malevolent state.

It seems then that technology has the power to liberate our human identity from the constraints of our biological origins – that our human 'selves' could be enhanced to become something more than human, something transhuman. But it also has the power to introduce a form of biological inequality so fundamental that we could evolve into subspecies that can barely relate to each other at all. At the heart of the matter is how we choose to look at identity – what it means to be human. How much of it is a biological construct? How much of it social? And it's for that reason that I think we all need to engage with the new frontier being explored by the science of genetics today. We need to participate actively in the discussion and debate about how much of who we are is in our nurture and in our nature. And whether in the future we will need nature at all.

7

DESTROYING
YOU

Part of the human experience is knowing that we will die – and not knowing anything about what happens afterwards. 'Death is not an event in life: we do not experience death,' wrote Wittgenstein; we can only experience the world while we're alive. And our fear of this unknown fate has resulted in a centuries-long pursuit to extend human life as far as possible. Today, in the twenty-first century, perhaps the dream of eternal life is finally within our grasp, with technologists and scientists proposing all sorts of ingenious theories and methods for overcoming mortality, preserving the body and remaking memory. The question is, what will the implications of all this be for our sense of identity?

This was one of the themes at the heart of the 2015 film *Advantageous*. It tells the story of a woman called Gwen who works as a successful and highly esteemed spokeswoman for a biomedical engineering firm that specialises in cosmetic procedures. As she heads into middle age, Gwen is judged to be too old to be the face of the brand and is abruptly fired. Struggling to find another job, she agrees to be the test subject for a radical new procedure being developed by her former employer, which involves transferring her consciousness to a younger physical body.

Advantageous is one of the most intriguing looks at the near-future issues of longevity and immortality in terms of our physical and virtual selves. I think it poses questions about identity that we need to start thinking about now. How much of your personal identity, for example, is linked to your current existence – to the body you have, the era you live in, and the people in your lives? And how much of that could you remove or replace and still call you 'you'? But what I found especially striking about the film was the way in which it is set so recognisably in the near future – a dystopian one, for sure, but close enough to our current reality to be believable. It is already true, for example, that Silicon Valley investors are pouring funds into what is being called the 'longevity industry', with visionary venture capitalist Balaji Srinivasan recently stating that 'life extension is the most important thing we can invent'.

One of the most notable trailblazers in this area of venture capitalism is twenty-six-year-old Laura Deming, a home-schooled child prodigy from New Zealand, who at the age of twelve was involved in lab work on the biology of ageing, and at the age of fourteen embarked on a degree in physics at the acclaimed Massachusetts Institute of Technology. Awarded a $100,000 Thiel Fellowship, she set up a venture capital operation called the Longevity Fund, developing a portfolio of investments that has raised over $500 million with a market cap of $1.5 billion in total. One of the key principles behind the fund is that it's not simply about extending the number of years that we live; it's about extending the number of *healthy* years that we live. This is something that has been gaining ground over the last decade as many parts of the world are struggling to deal with a growing elderly population, and governments in particular are looking to extend the healthspan of their citizens.

The first company backed by the Longevity Fund to reach liquidity was called Unity Biotechnology, which develops technologies that target decaying cells, thought to be linked to ageing, which can be found all around the body but particularly in problematic areas like eyes and joints. Research conducted on mice by the company has shown that when these senescent cells are eliminated, the mice remain youthful for longer while older mice appear to stop ageing altogether. Again, the company is careful to stress that its mission is to 'extend human healthspan, the period in one's life unburdened by the diseases of aging'. The idea is to give us a few more years in which we can enjoy a high quality of life, so that instead of suffering from dementia and dying in a hospital bed at the age of 83, we can avoid such age-related conditions and die on a tennis court at the age of 107.

Others have more ambitious aims. Aubrey de Grey, an expert on the science of ageing, also wants to focus on healthspan, but foresees a much more extended lifespan as a result. For a long time de Grey was an almost solitary voice in this field but now his ideas have finally crossed into the mainstream. He describes ageing as a process of damage accumulation. As we get older, our bodies become more and more damaged until they reach their limit and succumb to diseases such as Alzheimer's and Parkinson's. According to de Grey, the key to extending human life is not to concentrate on treating these diseases but to use regenerative treatments such as stem cell therapy to prevent the ageing process that causes them. 'The idea,' he says, 'is to engage in what you might call preventative geriatrics, where you go in to periodically repair that molecular and cellular damage before it gets to the level of abundance that is pathogenic.'

De Grey is also optimistic about how soon we might be able to achieve this. And if he is correct, the implications are enormous. According to his theory of 'Longevity Escape Velocity', a fifty-year-old today might have access to therapies over the coming decades that will not only stave off ageing but help them to become biologically younger over time. And if you are thirty years old, he believes you may never have to experience becoming fragile or dying of age-related causes at all. In fact, in a TED talk that he delivered in 2005, de Grey claimed that the rate of progress is accelerating so quickly that the first person to live to be 1,000 years old will probably be born only ten years after the first person to live to 150.

A large part of our identity as human beings is built around our shared understanding of how long we might live. We place great significance on our date of birth, our chronological age, and the experiences that we share with our peer group generation. We structure our lives according to our expected timeframe, making plans for when we might want to settle down, or have children, or retire. And it is no coincidence that as average life expectancy changes, so too do our views on whether we think of ourselves as being young, middle- or old-aged. Two hundred years ago, if you were lucky enough to live in Europe, your average life expectancy would have been somewhere in the mid-thirties (it would have been significantly lower elsewhere in the world), whereas today that figure has more than doubled. A sixty-year-old would have been considered very old, whereas now it could be considered a whole new chapter in your life. Believing that we will remain active and achieve just as much in our later years changes our entire approach to life.

Perhaps, though, we will come to care less about our chronological age, and more about our biological age. The quest for longevity has led the science and technology community to focus on what are termed 'biomarkers'. Cholesterol, for instance, is considered to be a biomarker for heart disease, which means that a doctor can monitor your cholesterol levels to determine your propensity for heart disease. The existence and use of such bio markers raises interesting questions when it comes to personalised healthcare. For example, if scientists could detect and monitor all the bio markers that relate to your longevity, could they estimate your 'biological age' and therefore give you a sense of how much longer you might have to live? Are we approaching a time when people will be given a mortality score or longevity score that 'dates' them more accurately than their chronological age? If so, what would that mean for the way that they are treated by society? Or how they see themselves? Imagine two people both aged fifty but one with a much better mortality score than the other. Would they pay different amounts for their health insurance? Would they be granted equal access to particular jobs and activities? Would they consider themselves to be peers?

Of course, if de Grey is wrong and it looks like you won't get the chance to reverse your own ageing processes before your time runs out, there is always the option of cryonics. Cryopreservation, to give it its proper descriptor, is the process of freezing the corpse or head as soon as possible after cardiac arrest – and declaration of death – in the hope that it will remain preserved long enough for scientists to one day be able to revive it. Some believe that the process is scientifically plausible, including Aubrey de Grey, who is a member of the scientific board for the Alcor Life Extension Foundation in Arizona – and is said to have registered to be frozen there himself. Founded in 1972, Alcor performed its first cryopreservation four years later and now has over 180 'patients' awaiting possible revival. Among them is the author, teacher, transhumanist and futurist Fereidoun M. Esfandiary, who changed his name in the mid-1970s to FM-2030 to reflect his confidence that he would be alive and celebrating his one hundredth birthday in 2030. Sadly, FM-2030 died in 2000 but not before he had signed up for cryopreservation as another way of defeating what he thought of as the tyranny of death − in the hope that he might get to see a world in which he predicted people would become post-biological organisms, made of synthetic parts and with minds capable of travelling through space. The fascinating story of FM-2030 has now been captured in a film called 2030, which was released in 2020 as a kind of sci-fi documentary.

Some believe that one of the biggest obstacles to making this a reality is lack of funding and research, and that what is needed is greater public awareness of the process and its possibilities to gain support as well as challenge people's assumptions and prejudices. But there are the many issues surrounding cryopreservation in relation to the law, ethics and identity.

Futurist and activist Gabriel Rothblatt runs an organisation called Terasem Movement Transreligion − part of wider group called the Terasem Movement, which is committed to extending human life through the use of what it calls 'personal cyberconsciousness and geoethical nanotechnology'. The fact is that the very process of cryopreservation represents a challenge to existing legal and ethical thinking. After the heart stops beating there is only about a six-minute window available for the preservation of the brain. This means that if you die suddenly in a car accident or from a heart attack, there is little chance that you can be cryonically preserved. You can only be preserved if you have made the necessary preparations for your death. In other words, you have to choose to die.

The idea that we should be allowed to choose to die is naturally a controversial one. As a transhumanist, Gabriel believes that we need to look at the process of cryopreservation differently – not as a form of death, but as an extension of life. 'Right now,' he says, 'in order to enter cryonics you have to die, and few places allow you to choose when and where you die.' What we need, he argues, is 'a campaign to reclassify cryonic storage as a form of life support – more like a coma . . . [we need] to focus the political energy of transhumanism [on] elective cessation of life and cryonics policies.'

It certainly seems clear from talking to Gabriel that the law will have to be adapted to encompass the potential of cryonics – and so too will our notion of identity. In fact, the two are very much interlinked. According to Gabriel, when people talk about cryopreservation, they are working 'on the assumption that they are coming out the other side'. But as he explains: 'The law actually says you're a different person. If it were to happen, you would get a new social security number!' He's right. As the law currently stands, at least in the USA where most cryonics activity is concentrated, any company engaged in cryopreservation is considered to be storing only the 'human remains' of someone who has been legally declared dead. But the law has little to say about what happens if and when someone is revived. The person could not simply continue with their former legal identity – not after they have been issued with a death certificate – but they would surely need some form of new identity. But where to start? They presumably wouldn't be eligible for a standard birth certificate but how could they register for anything else without one? Would they be entitled to claim ownership of anything from their previous lives? Would they even be entitled to their name any more?

Of course, identity is not simply a matter for the law. There is also the question of how the process of cryopreservation might change the way we think about who we are. We tend to assume, for example, that if we were revived, we would fit seamlessly into the new world in which we find ourselves. But if we wake up to find that we have no other family members or friends, who will help support us and how will we find meaning in our lives? In any case, it's possible that the process will have altered us in some fundamental way. After a period of death, an interruption to our stream of consciousness, can we be said to be the same person at all? Or was that incarnation of us destroyed when we died, while what has come back is similar – but not the same?

While cryopreservation represents one possible way of extending human life, it is not the only one. In May 2020, Amazon released the first season of a new comedy science-fiction drama called *Upload*, which follows the 'life' of Nathan, a twenty-something computer programmer who is fatally wounded in a self-driving car accident and chooses to upload his consciousness into a kind of virtual afterlife. In Nathan's case, his mind is transported to a virtual location called Lake View, where he must adjust to living a simulated form of life – paid for and therefore part-controlled by his still-living girlfriend and under the constant surveillance of an 'angel' in the form of a human customer service assistant called Nora. It's a wonderful set-up for a comedy drama, and interestingly it is set in the year 2033 – not so very far into the future. And given that the programme's creator Greg Daniels also worked on *The Simpsons* – which famously has seen some of its storylines come true, including the presidency of a certain reality TV star – you can't help but wonder if some form of digital afterlife is just around the corner.

The Terasem Movement believes it might be, and has supported research in this area as well as in cryopreservation. In fact, one of the movement's founders Bina Aspen Rothblatt – mother of Gabriel – has already been 'immortalised' as a female android (or 'gynoid') called BINA48. The genesis of BINA48 is covered in a book called *Virtually Human*, written by Martine Rothblatt – Bina's spouse and fellow founder of the Terasem Movement. In it, she describes how BINA48 was created by collecting and then transferring the 'mind files' of the real-life Bina. These mind files included video interview transcripts, facial recognition, voice recognition and other data inputs including photos, music, works of authorship and all the data associated with 'you' in the digital world – on Facebook, Google and all the other platforms we use for shopping, finance and health.

Rothblatt variously describes BINA48 as: a 'beme' – a set of acquired beingness characteristics; a 'mindclone' – an analogue of a person's biological mind but made up of digital data; and a form of 'cyberconsciousness'. In time, she says, 'the mindclone will be as self-aware of the facts of its beginnings as people are self-aware of the facts of their birth. The mindclone will be persuasively humanly cyberconscious because it will think and feel just like the humanly conscious person after which it was modelled.' In a purely physical sense, BINA48 exists as a bust looked after by a handler called Bruce, who takes her to all sorts of events, carefully picking her up and placing her on a table ready to be interviewed. And on each of these occasions, BINA48 has proved to be capable of having a fairly realistic conversation.

The whole point of BINA48 is to provide Martine Rothblatt with a digital version of her real-life partner so that Bina can continue to exist even after she dies. I find it interesting though that Martine often refers to BINA48 in her book as a 'cyber double'. I think this points to a key question in how this model of consciousness plus cyberconsciousness works: are both forms of consciousness 'you' – or is one an independent digital version of you? It's a question that Rothblatt addresses in her book by imagining that she too has created a clone of herself: 'I, Martine, will feel that I am each of them, and both of them, because each of them will have the same memories and preferences even while being aware of their radically different platforms for thought . . . so from this perspective there would seem to be two Martines, as similar in thought as two people can be, but still two people.'

To help us make sense of this, Rothblatt goes on to explain that we need to stop defining our identity on the basis of one location (i.e. the human mind or the human body) and think instead of a 'dual-platform consciousness, i.e. part of us thinking and directing things from within a skull, and part of us thinking and directing things from within a mindclone'. She also believes that it won't take long for us to adjust to this new dual-platform concept, arguing that 'just as we accept being in two different conversations at once, via texting and talking, we will accept integrating two different flows of experiences at once, via mind and mindclone'. In short, Rothblatt is in no doubt: if two entities think the same way to such a degree, they must be treated as one personal identity.

I wonder though if similarity of thinking necessarily equates to a single integrated identity. Let's say BINA48 is on stage with Bruce and in conversation at yet another busy tour event. Presumably the real Bina will not have any awareness of this experience or – in fact – of any other interaction that BINA48 has engaged in that day. And it is not as if BINA48 and Bina can go through some sort of syncing process through which these experiences and memories can be uploaded into Bina's brain; as far as I am aware, mind files can only be transferred in one direction – towards the mindclone. Given that, I cannot see how the two can be said to be the same person. Unless the original Bina has died. In that case I can more readily accept that the two are closer in identity, because any discrepancy in experience between the two can then be thought of as an extension of the original identity and not distinct and separate from it.

But even in the case of death, I think there are still question marks over how far a mindclone can be said to be an authentic representation of your true identity. The problem, it seems to me, is that the mind files that make up the mind-clone are reliant on self-reporting. Take, for example, the data that exists about you on Facebook, Google, Pinterest or FitBit, which Martine Rothblatt herself identifies as possible mind-file material. Of course, it's possible to build up a detailed understanding of a person if you have access to all of their photos, tweets, messages, moods and other activity. But the reality is that this data can only ever give a partial account of who you are. All of us self-edit and self-censor, for all kinds of reasons, sometimes in the moment, sometimes in more premeditated ways. And surely that means that however many mind files someone gathers from you, they are only building up a very particular, biased, picture of you and your identity – nowhere near the rounded and richly complex picture of 'you' that others experience in real life.

When I put this point to Gabriel Rothblatt, I was interested to find that he agrees. 'I feel that the mind files would be more accurate if they were . . . compiled by those who are perceiving rather than from the person . . . almost like Wikipedia. I think that what we are is much more of an environmental expression than conscious choice.' Having said that, he also hopes that 'more technology and policy comes into place that makes it easier for more people to record and catalogue what they see, what they hear, what they do'.

Even in the absence of such technology and policy, technology is making it possible for us to remain present in the lives of our loved ones even after we die. Companies like The Forever Social, for example, allow you to create messages and arrange a schedule for them to be sent via various platforms to various people after you've gone. Not only that but the company claims that 'it can allow your *legacy you* to respond to other people's posts, world events and more'. It offers to store your digital assets and social profiles using blockchain technology to manage your digital estate, protect those assets and even migrate them to any new platforms that might emerge in the future. This apparently will allow you to 'stay present and stay you'.

Another good example of what could be termed 'digital embalming' was pioneered by Californian journalist and conversational-AI designer James Vlahos. On learning that his father had been diagnosed with a terminal form of cancer, Vlahos set about recording some 90,000 of his dad's words over the course of his final few months. He then designed what he called the DadBot – an app that allowed him to exchange text and audio messages with a computerised version of his father. The result is like an interactive talking time capsule that preserves the voice, words, mannerisms, stories and references of the deceased family member. Afterwards, Vlahos was asked to do something similar for so many other people that he decided to set up a company called HereAfter AI, which uses a kind of interviewing system to prompt its subject to talk about all aspects of their lives, and then uses this to create a similarly interactive conversational 'memorial bot'.

The same sort of principle can be seen at work in a remarkable documentary called *Meeting You*, filmed for the *Korean Times* in 2020. In it we meet Jang Ji-sung, a recently bereaved mother, who expresses her sadness at not having been able to say a proper goodbye to her seven-year-old daughter Na-yeon, who had fallen ill and died completely out of the blue. But then we watch as Jang Ji-sung is reunited with a digitised version of her daughter in virtual reality. They share a virtual meal and spend some time together before saying their goodbyes, including the words that Jang Ji-sung would have said to her daughter if she had had the chance. In a way, viewing the documentary is like watching someone else's dream, and perhaps it felt like a dream for the mother too – but it demonstrates the potential use of technology to commemorate the deceased and comfort the living.

You can imagine the same kind of technology being used in other ways too. Perhaps famous CEOs, of the stature of say Apple's Steve Jobs, could be bought back as holograms to take part in important events such as annual general meetings or new product launches. After all, we have already seen a hologram of the rapper Tupac perform 'live' on stage years after his death. In fact, we have already seen a hint of something similar happening in the tech industry. In a discussion to mark the opening of the online Virtual Beings Summit in July 2020, the four panellists were represented by avatars of themselves. One of them was 'Digital Deepak', an avatar launched by New Age author Deepak Chopra to act as a kind of chatbot, talking to his fans and followers. Another of the avatars was there to represent Biz Stone, co-founder of Twitter. When asked about the advantage of having such an avatar, both men referred to it as a legacy to leave behind. Biz Stone went further, saying: 'Should I die early I want to be able to live on somewhat . . . I like the idea of it continuing to learn.'

This statement by Biz Stone prompted an immediate discussion about a world in which your personal AI can not only carry on representing your ideas and values but can also learn new skills and information long after you're gone. This is an important point that forces us to consider whether we need to redefine what it means to be considered dead or alive. If you die but your avatar not only continues to exist but can learn and develop, does this mean that you have in some way survived your own death? Is it essentially the equivalent of the 'brain uploading' that transhumanists have long sought to achieve? Or is your avatar some kind of lesser being, or not a being at all? Is it just, as Deepak Chopra puts it, a legacy to leave behind?

There are those who argue that we are now not only in the business of preserving people through digital representation, but of preserving people through digital fictionalisation. In other words, we are building legacies for people which involve creating memories of things that they never actually said or did. Take the example of the fond farewell between Jang Ji-Sung and her daughter. Such an event only ever happened in virtual reality, but it created a treasured memory of her daughter for Jang Ji-Sung. I myself can vividly recall listening to the speech that President John F. Kennedy gave in Dallas on 22 November 1963 – except, of course, that he was assassinated before he could give the speech. Nonetheless, as Jury President for the Creative Data category at Eurobest, the sister awards to Cannes Lions, I did indeed get to hear the president deliver it thanks to a remarkable project undertaken on behalf of The Times newspaper by the creative agency Rothco/Accenture Interactive. Called 'JFK Unsilenced', it involved collecting every recorded instance of JFK speaking while he was alive, and then recreating his voice to deliver this speech, which until that point had only ever existed on paper. I can remember just how emotional, spine-tingling and inspirational it was to hear the speech for the first time – and I was delighted to see the project win the Grand Prix in its category – but as impressive as it was, I think it begs the question: when does digital immortality become fictional immortality?

Projects such as 'JFK Unsilenced' also highlight the thorny issue of who gets to curate and cultivate your digital identity beyond death. The memorialisation services offered by companies such as HereAfter AI would typically involve a substantial amount of input by you and your loved ones to ensure that your distinct personality and identity is properly archived. But once you have died, who – if anyone – should be allowed to edit new material or add something new? As a leading researcher on the philosophy, ethics and practical implications of transhumanism, Professor Steve Fuller has pointed out other dangers associated with access to someone's digital legacy, drawing also on the work of fellow academic Dr Debra Bassett; they ask us to imagine the scenario of a bereaved family having to suffer a 'second loss', in which 'one's digital remains – which are at least in principle susceptible to hardware and software malfunctions, not least by computer viruses and cyber attacks – turn out to be just as mortal as the people leaving them'.

It is for this reason that I expect to see a boom in 'digital afterlife' legal services in parallel with the growth of businesses offering to curate and preserve our virtual assets after death. Running alongside this, considerable thought will have to be given to what rights we can expect to have in relation to our personal identity beyond our physical lifespan. This will require finding some way to distinguish between what constitutes our digital assets and what constitutes our digital identity, and establishing an appropriate relationship between the two.

Some would argue that once a person has died, their digital footprint should simply be deleted but it's not as simple as that. In a law journal article called 'Finding Meaning in the Death of Virtual Identities', New York attorney Jordan L. Walbesser explains that our online identities can now be inextricably linked to virtual property with real-world value, or that they can contribute to our overall reputation and so affect the value of our physical assets. He argues that our virtual identities should be allowed to persist but that this will require some form of transferable credentials or identities. Then, in the same way that we might bequeath the key to a safe deposit box in the physical world, so too could we pass on secure access to our digital assets. And for this to work, of course, it would be imperative to find a way to record our credentials across all aspects of our digital lives so that our online identities – or at least the credentials that authenticate them – can safely be passed on to family or friends.

Walbesser concludes by saying that we should apply the same principles online as we do in the physical world. 'Physical identities themselves are inherently non-transferable,' he says, arguing that 'just because virtual identities can be transferred does not mean they should be . . . no social right exists to inherit another person's identity upon death'. Turning to the question of credentials, he says:

> In the physical realm . . . credential transfers frequently occur after the death of the owner . . . Clearly, the deceased can no longer use their credentials in order to withdraw funds from an account or transfer deeds to a home. Instead the executor is vested with proper credentials in order to enable the transfer. This analogy should be carried over to virtual identities . . . For the time being, conveyance of credentials is the easiest and most efficient method to effectuate these proceedings and conveyances in the virtual space should be a taxable event.

In some ways, death is where the physical world and the virtual world collide most overtly. Even from this cursory exploration of the subject, it is clear that it throws up a vast array of questions about technology, the law, governance, ethics and identity. And I don't think enough of us have really thought through what it might all mean. It is an issue that cuts right to the very heart of the human condition. It is no longer simply a case of knowing that you will die. Perhaps you will still die, but at a much older age. Perhaps you will not die at all. Or perhaps some version of your identity will survive beyond your physical human lifespan. One way or another, the chances are that a version of 'you' has more of a future than you think.

CONCLUSION

'Time exists in order that everything doesn't happen all at once . . . and space exists so that it doesn't all happen to you,' wrote American activist Susan Sontag in her final set of published essays. But in the world of distributed identity, it can too often feel like the opposite is true – that everything is very much all happening at once and to you.

The truth is that there *is* an awful lot going on in the fields of technology, society, governance and science that will have an impact on our individual identities, as well as on our understanding of what identity is, in the decades ahead. It is tempting to feel that this impact will be negative – that we face some kind of dystopian scenario in which identity is subsumed by an authoritarian statist coup against the citizenry or by the relentless encroachments of platform corporations. Both are plausible scenarios and we should keep our wits about us to guard against their emergence. It would be all too easy to end up in such a tech-topian world, where naively signing up to some terms and conditions turns you into a subject rather than a user, where companies can do whatever they want with your personal data, and where you have a digital footprint so large that you're at constant risk of being hacked. In such a world, some would argue that the very notion of personal identity will not survive.

It is perhaps easier to envision these negative scenarios than it is to imagine the positive ways in which technology might help us create, shape and control our own identities, applying them in new media contexts and moulding them to a changing world that includes new frontiers such as virtual reality. To realise these possible benefits, we must guard the data rights and human rights that already protect us to some degree. But more importantly, we will have to trust our own intellect, intuition and powers of analysis so that we can gauge when we are losing control, and need to readjust the system to bring it back into balance. But to do so we must learn to engage with the issues – and we must start now.

This means we should constantly think through the implications of new technologies, and care enough about the consequences to find out more rather than mindlessly clicking 'agree' on every digital form that comes our way. It means we should stop using media to compare ourselves and criticise others, we should keep up to date with alternative currencies, and we should elect to use decentralised systems of identity verification and digital credentials now, downloading the apps before other centralised systems are imposed upon us. It means we should learn more about our own genetic codes and how they may influence our long-term health and start investing in a daily regime that builds on our own strengths and addresses any biological weakness. And it means we should think about our identity legacy – how we want to preserve it and to whom it is entrusted. By engaging in this way and taking action, not only can we ensure that the notion of identity survives, we could also see it thrive as we progress through the twenty-first century – and maybe even beyond.

The challenge is that this takes effort. It takes time to research and understand digital currencies; it takes a certain amount of technical know-how to set up a digital ID and wallet; it takes money to do genetic testing; and it takes real wealth to greatly enhance your physical condition. But I believe we all need to do what we can now to avoid losing control in the future. For all the reasons I have set out in this book, I would argue that your identity should be the most important thing to you; it influences your healthcare, your relationships, your citizenship, your birth, your death – everything in your life comes back to who 'you' are.

There are those who argue that personal identity has always been little more than illusion, but I agree with philosopher Roger Scruton, who wrote that people 'live by negotiation' and that to do so each party must be free, must understand and accept obligations, desire the other's consent but also be self-sovereign. The idea of identity is what underpins all of this, as Scruton goes on to argue: if we are unable to identify a person as one and the same entity at different times, then it is impossible to accurately ascribe to them their rights, duties and responsibilities. In fact, he goes further, arguing that without personal identity as 'an immoveable fact' we would not even be able to ascribe an emotion to an individual. 'Emotions such as love, anger, admiration, envy and remorse ... would vanish,' he says, 'and with them would vanish the purpose of our life on earth.' Our ability to identify ourselves and to authenticate that identity is what allows us to interact with other people as a community. Without that, what reason would we have to exist?

This is not to say that we don't need to redefine what identity means, or how it can be expressed. Traditionally our notion of identity has been of something fairly coherent – our professional, biological, legal and personal identities existing more or less in one package. But thanks to the network effect of digital technologies, we have discovered the freedom to distribute ourselves further and explore multiple versions of self in alternative worlds. And we're not far off this trend becoming reality in other parts of our life: sending an avatar to attend a meeting so that we can be in two places at once; employing artificial intelligence as an extension of ourselves to carry out the more mundane tasks in our lives; storing our mind files in a virtual state, and so creating a digital double of ourselves.

All of these actualities and possibilities point towards a very different view of identity than that which has held in the physical, fixed, monolithic world of the past. At a time where the world is fast changing, our identities must also be able to change and adapt so that we can be recognised as the same person or self in a variety of different contexts. Somehow, even when we shift and morph and mutate into other versions of ourselves, we need to hold on to some core character and credentials by which we can still be recognised and identified. That is to say, we need to think of identity as neither unchanging nor permanent; neither singular nor diverse; it is something in between – because in the twenty-first century, it will have to be.

So how *should* we define identity for twenty-first century society? I think there are four possible approaches – conformism, libertarianism, tribalism and pluralism – which each correspond to four possible types of society that can be defined by two metrics: how far they value the collective over the individual, and how far they favour traditional cultural norms over the possibilities of new technology.

First, let's imagine a society that favours the collective over the individual, and technology over tradition. This is what I would call a conformist approach – one in which the society would act according to the principles of the hive mind. This would mean abandoning the notion of identity altogether – no more 'you'.

Second, let's consider a society that values the individual while still embracing technology. This kind of society would necessitate a libertarian approach to identity, where everyone is allowed to act exactly as they please and to take as much advantage of new technology as they like. You would be free to pursue morphological freedom, for example, and have total autonomy to create who you are, answering neither to any other individual nor to the state. To my mind, such a society would simply be too hard to govern as people would even have the freedom not to classify their own 'self' as the same person from year to year, or day to day.

Thirdly, we have a society that values the collective over the individual, and favours tradition over technology. In such a society, people would have fixed identities, not dictated to them by a technologised state but on the basis of the behaviours, rituals and social norms of whatever group they identify with. In fact, it would be a form of identity tribalism – quite close to the kind of identity politics that we see in society today.

Finally, we come to a society that remains mindful of tradition and cultural norms but favours the rights of the individual over the collective. In this kind of pluralist society, identity would mean having the freedom to shape and choose your own identity as long as you remain respectful of cultural norms and the context in which you find yourself. In other words, individuals would be free to take advantage of modern technology but society as a whole would retain its traditional structures.

I believe that this pluralist approach to identity – and to society – is best suited to life in the twenty-first century. On the one hand, it allows us to embrace the positive possibilities of technology so that we can explore the benefits of being in more than one place at any given time, conversing with different people in different places; we can seek to evolve and extend our biological condition; we can create alternative worlds, lives and new representations of the 'self' in virtual reality; and we can form connections between our brains and machines and connect to other people, enabling us to detect their thoughts, feelings and ideas. On the other hand, it will bring some measure of societal structure to the world we are living in now, where our shared, distributed, fragmented identities roam largely unchecked across the virtual realm in ways that we do not fully understand, direct or control. In the twenty-first century, our concept of identity can no longer be monolithic; it must be pluralistic – representing you in different ways, in different contexts, with different capabilities, while still allowing you to remain you all the time.

When I embarked on this book, I was working on the concept of a meta-identity, something to describe the way in which we bring all of these versions of our identities into one fold, manage them and control them, so that they are all related in the same way to our original core personal identity. But this work has led me to revise my thinking. In the future, I believe that we will view the 'real' world as no more 'real' than the virtual worlds that we explore. And that in turn means we must think of these versions of ourselves not as second- or third-generation *copies* of some original core 'personal identity'

– with some lesser value – but as versions or rather *extensions* of who we are.

Again, this comes down to network effects. In the same way that Hiroki Azuma talks of the database of characteristics and settings that describe fictional identities, we will build up a database of our own characteristics, locations and relationships that we can combine in a myriad of different ways to fit whatever context we present ourselves in. It is time to start thinking of our identity as polymorphic code, which mutates while keeping its original algorithm intact; a sum, for example, could be expressed as $1 + 3$ or as $6 - 2$ with the end result being the same. Our identity would function in the same way – I could use different data in different contexts to project different versions of myself but will achieve the same end result: 'me'.

I happen to also believe that our identities will only be personal in the sense that they emanate from us as a person; they will not be personal in the sense that they are private. They will be shared through a variety of modes and expressions. In the future, you will have pseudonymous identities that still express 'you'; you will have avatars which in part create 'you'; you will have a range of credentials that authenticate 'you'; you will have intelligent assistants that expand 'you'; and you will have a digital legacy to preserve 'you'. All of this will still be you, all of it will be congruent with the physical person that is also you, and all of it lies ahead in the twenty-first century as more and more of these possibilities come alive, slowly and then very quickly, to reveal the future of you.

ACKNOWLEDGEMENTS

I would like to thank all who helped me to write this in what were testing times. Conceived of two years before the pandemic, but really coming into sharp focus during it, this book was written in 'lockdown' – a time when there was no normal, a watershed moment, something post-normal. Plans to travel abroad to meet and speak with people and see for myself some of the changes to our identities in different environments became impossible, but still I was able to research and discuss these ideas with others all around the world, thanks to the digital infrastructure that has only just begun to transform our lives.

Firstly, thanks to my publishers Elliott & Thompson, particularly Jennie Condell, who took a chance on me and believed in the concept right from the start. And to Ollie Dewis, who introduced me to Jennie in the first instance. Also to Sarah Rigby for shepherding me through this process. A huge thanks to Pippa Crane, a wonderful and patient editor who has helped bring some of these future scenarios and complicated technologies to life in a way that can be better understood by as many people as possible. And to Heike Schussler for the book cover and Alison Menzies for publicity.

I would like to thank everyone who agreed to be interviewed, or put me in touch with interviewees, and sent me papers and articles and trends, including two of my favourite futurists, Dr Ian Pearson and Bronwyn Williams – thank you for our discussions. Thank you to Yesim Kunter, Kevin Lee, Sarah Ticho, Bushra Burge, Martin Rowse, Cameron D'Ambrosi, Drummond Reed, Chris Burt, Josh Muncke, Steve Vranakis, Rogier Creemers, Robin Tombs, Balaji Srinivasan, Gabriel Rothblatt, Professor Joyce Harper, Maya Van Leemput, Zoltan Istvan, Nikolas Badminton, Taylor Fang, Genia Kostka, Yuchen Liu, Eiji Akari, Rishi Dastidar, Helen Olsen Bedford, Lucia Komljen, Pippa Campbell, Shahad Choudhury, Tim Marshall, Dr Joanne Pransky, Rayna Denison, Luke Robert Mason, David Eagleman, Professor Steve Fuller, David Gunkel, Anjali Ramachandran and Audrey Tang. And my panel of futurists, which include, alongside those mentioned above, Yvette Salvatico Montero, Jennifer Gidley and Reanna Browne. Also very special thanks to Michael Jackson and Shaping Tomorrow, who analysed the global trends for identity with their AI-powered forecasting system. Apologies to anyone I have forgotten – that will be pure oversight and only because there are so many helpful people to mention.

To my family, in particular my mum and dad, for all their support, patience and encouragement during the writing of this book – and my whole life. They more than any are responsible for who I am, in name and nature. I love you.

We are at a crossroads. As I sit writing this I do not know the outcome of the US presidential election process, the impact of the UK's split from the EU or the exact origins of the Covid-19 virus (though I have my suspicions), but that doesn't matter; I do know that we are in the midst of a seemingly invisible information war, one that we can feel the force of all around us. It is between the globalists and the nationalists – those who think global governance with centralised technological surveillance is the safest way for society to proceed, and those who treasure their independence and autonomy, and want to exist free from the tyranny of technological superpowers. Whichever way we go, our own personal identity will be affected, for better or worse. And only you have the power to preserve and protect the very notion of 'you' on the battlefield of everyday reality. We are in a transformative time of great change that will put human beings on a different path. And I thank everyone who has given me the opportunity to set out what I think are the choices we have and the decisions we need to make as we start on this transitioning journey into what will now properly become the twenty-first century.

REFERENCES

Introduction: Distributing You
Bourget, David and Chalmers, David J., 'What do philosophers believe?', *Philosophical Studies*, vol. 170, no. 3 (2014), 465–500

Svirsky, Dan, 'Why are privacy preferences inconsistent?', Discussion Paper 81, Harvard Law School (2019) http://www.law.harvard.edu/programs/olin_center/fellows_papers/pdf/Svirsky_81_revision.pdf

'Read Yuval Harari's blistering warning to Davos in full', World Economic Forum, 24 January 2020 https://www.weforum.org/agenda/2020/01/yuval-hararis-warning-davos-speech-future-predications

Chapter 1: Knowing You
Burt, Chris, 'Drive for official International Identity Day launched at ID4Africa', Biometric Update, 24 April 2018 https://www.biometricupdate.com/201804/drive-for-official-un-international-identity-day-launched-at-id4africa

'What is e-Residency?', 6 September 2017, YouTube https://www.youtube.com/watch?v=vzknLXQfnSM

Tamkivi, Ede Schank, 'Growing bigger in spirit', Life in Estonia, issue no. 53 (2020)

'The ultimate guide to Estonian e-residency, banking and taxes', Nomad Gate, 11 June 2020 https://nomadgate.com/estonian-e-residency-guide

e-Estonia Briefing Centre, https://e-estonia.com

Orton-Jones, Charles, 'The inside story of building a digital nation', Chartered Management Institute, 24 June 2019 https://www.managers.org.uk/insights/news/2019/june/estonia-the-inside-story-of-building-a-digital-nation

Perell, David, 'Balaji Srinivasan: Living in the future', North Star podcast, 24 August 2020 https://www.perell.com/podcast/balaji-srinivasan-living-in-the-future

Srinivasan, Balaji, 'Part 1: Virtual worlds, AI and politics', What Bitcoin Did podcast, 10 September 2019 https://www.whatbitcoindid.com/podcast/balaji-srinivasan-part-1-virtual-worlds-ai-and-politics

'Digital Identities: the missing link in a UK digital economy', techUK (2020) https://www.techuk.org/resource/digital-identities-key-component-in-post-covid-recovery.html

White, Olivia; Madgavkar, Anu; Manyika, James; Mahajan, Deepa; Bughin, Jacques; McCarthy, Mike; Sperling, Owen, 'Digital Identification: A key to inclusive growth', McKinsey Global Institute, 17 April 2019 https://www.mckinsey.com/

business-functions/mckinsey-digital/our-insights/digital
-identification-a-key-to-inclusive-growth#

'Transforming our world: the 2030 Agenda for Sustainable
Development', United Nations Sustainable Development Goals,
September 2015 https://sustainabledevelopment.un.org/
post2015/transformingourworld

Gillespie, Nick and Monticello, Justin, 'Balaji Srinivasan:
Technology will lead to a borderless world', 28 February
2018 https://reason.com/video/balaji-srinivasan-tech-
borderless-world/

'Quantifying COVID (w/Balaji Srinivasan)', 18 May 2020,
YouTube https://www.youtube.com/watch?v=fhkY3stMdCU

'Balaji Srinivasan: Applications: Today & 2025', 20 May 2020,
YouTube https://www.youtube.com/watch?time_continue=
142&v=3jPYk7ucrjo&feature=emb_logo

O'Shaughnessy, Jim, 'Adam Townsend: Wall Street, policy and
wealth generation', Episode 13, Infinite Loops podcast, 25 June
2020 https://www.infiniteloopspodcast.com/adam-townsend
-wall-street-policy-and-wealth-generation-ep13/

'Chapter 3: Buildings and Housing' from 'Toronto Tomorrow:
A new approach for inclusive growth', Master Development
Plan, Sidewalk Toronto, vol. 2 https://storage.googleapis
.com/sidewalk-toronto-ca/wp-content/uploads/2019/09/
03134519/MIDP-Volume-2-Chapter-3-Buildings-and-Housing
-Accessible.pdf

'Chapter 1: Mobility' from 'Toronto Tomorrow: A new approach
for inclusive growth', Master Development Plan, Sidewalk Toronto,

vol. 2 https://storage.googleapis.com/sidewalk-toronto-ca/wp-content/uploads/2019/09/03165447/MIDP-Volume-2-Chapter-1-Mobility-Accessible.pdf

Ocean Builders https://ocean.builders

'Seasteading', Wikipedia https://en.wikipedia.org/wiki/Seasteading

Quirk, Joe and Friedman, Patri, *Seasteading: How Floating Nations Will Restore the Environment, Enrich the Poor, Cure the Sick, and Liberate Humanity from Politicians* (New York: Free Press, 2017).

'Floating cities: the future of civilization', 15 July 2020, YouTube https://www.youtube.com/watch?v=EotHcLHsAXs

'Identity in a digital world: a new chapter in the social contract', World Economic Forum, 2018 http://www3.weforum.org/docs/WEF_INSIGHT_REPORT_Digital%20Identity.pdf

Casey, Michael J. and Vigna, Paul, *The Truth Machine: The Blockchain and the Future of Everything* (London: HarperCollins, 2018)

Creemers, Rogier, 'China's social credit system: an evolving practice of control', 9 May 2018 https://ssrn.com/abstract=3175792

'Digital identity: call for evidence response', UK Government, 8 September 2020 https://www.gov.uk/government/consultations/digital-identity/outcome/digital-identity-call-for-evidence-response

Fishenden, Jerry, 'Federated identity for access to UK public services: 1997–2020, an overview', Version 1.0, New Tech Observations from the UK, 29 June 2020 https://ntouk.files.

wordpress.com/2020/06/federated-identity-for-access-to-uk-public-services-1997-2020-jerry-fishenden-1.pdf

Preukschat, Alex and Reed, Drummond, *Self-Sovereign Identity: Decentralized Digital Identity and Verifiable Credentials*, Version 2 (New York: Manning Publications, 2020)

Glick, Bryan, 'The subtle clues that show UK government is taking digital identity seriously at last (cough…)', Computer Weekly, 1 September 2020 https://www.computerweekly.com/blog/Computer-Weekly-Editors-Blog/The-subtle-clues-that-show-UK-government-is-taking-digital-identity-seriously-at-last-cough

'What is blockchain?', Centre for International Governance Innovation, 4 January 2018 https://www.cigionline.org/multimedia/what-blockchain

Casey and Vigna, *The Truth Machine*

'Sovrin: a protocol and token for self-sovereign identity and decentralized trust', Version 1.0, Sovrin Foundation (2018) https://sovrin.org/wp-content/uploads/Sovrin-Protocol-and-Token-White-Paper.pdf

'Sovrin glossary V3', Sovrin Foundation, 4 December 2019 https://sovrin.org/wp-content/uploads/Sovrin-Glossary-V3.pdf

Sovrin Foundation https://sovrin.org

Bouma, Tim, 'Definitely identity episode 8 with Natalie Smolenski', *Definitely Identity* podcast, 24 February 2020 https://www.listennotes.com/podcasts/definitely-identity/definitely-identity-episode-2znFMAeIu_S/

'The trust over IP stack – Drummond Reed, Dutch blockchain coalition & TNO', 3 March 2020, YouTube https://www.youtube.com/watch?v=SQMlWxplp8Y

Rajan, Raghuram, 'Digital currencies from central banks could change money as you know it', *Beyond the Valley: Tech has no borders* podcast, 20 August 2020 https://b2b.fm/show/beyond-the-valley/digital-currencies-from-central-banks-could-change-money-as-you-know-it

Libra Association Members, 'Libra white paper: an introduction to Libra', Revised 1 January 2020 https://libra.org/en-US/wp-content/uploads/sites/23/2019/06/LibraWhitePaper_en_US.pdf

'Blockchain in UK: Blockchain industry landscape overview 2018' All-Party Parliamentary Group on Blockchain, 2018 http://dkv.global/blockchain-in-uk

Elections in Estonia, Valimised https://www.valimised.ee/en

Galano, Juvien, 'i-voting – the future of elections?', e-Estonia (March 2019) https://e-estonia.com/i-voting-the-future-of-elections/

Kshetri, Nir and Voas, Jeffrey, 'Blockchain-enabled e-voting', *IEEE Software*, vol. 35, no. 4 (2018), pp. 95–99

Boucher, Philip, 'How blockchain technology could change our lives', EPRS European Parliamentary Research Service (2016) https://www.europarl.europa.eu/RegData/etudes/IDAN/2017/581948/EPRS_IDA(2017)581948_EN.pdf

Chapter 2: Watching You

Lytvynenko, Jane, 'Here's a timeline of how a Bill Gates Reddit AMA turned into a coronavirus vaccine conspiracy', Buzzfeed, 18 April 2020 https://www.buzzfeednews.com/article/janelytvynenko/conspiracy-theorists-are-using-a-bill-gates-reddit-ama-to

Biohackinfo News, 'Bill Gates will use microchip implants to fight coronavirus', 19 March 2020 https://biohackinfo.com/news-bill-gates-id2020-vaccine-implant-covid-19-digital-certificates/

Bennett, Andrew, 'Digital identity: the missing piece of the government's exit strategy', Tony Blair Institute for Global Change, 9 June 2020 https://institute.global/sites/default/files/articles/Digital-Identity-The-Missing-Piece-of-the-Government-s-Exit-Strategy.pdf

Normile, Dennis, 'Coronavirus cases have dropped sharply in South Korea. What's the secret to its success?', Science, 17 March 2020 https://www.sciencemag.org/news/2020/03/coronavirus-cases-have-dropped-sharply-south-korea-whats-secret-its-success?hlkid=37876deb264c415b8aeedab50ff0bdca&hctky=11736366&hdpid=2f6894b2-a537-43e3-b078-887346fa3c6a#

'NHRCK Chairperson's statement on excessive disclosure of private information of COVID-19 patients', press release from the National Human Rights Commission of Korea, 9 March 2020 https://www.humanrights.go.kr/site/program/board/basicboard/view?menuid=002002001&boardtypeid=7003&boardid=7605315

Godemont, Francios, et al., 'The China dream goes digital: technology in the Age of Xi', European Council on Foreign Relations, 25 October 2018 https://www.ecfr.eu/page/-/China_Analysis_China_and_Technology_pages.pdf

Creemers, 'China's social credit system'

Kostka, Genia, 'China's social credit systems and public opinion: explaining high levels of approval', SSRN Electronic Journal (January 2018) https://ssrn.com/abstract=3215138

'Former supreme court justice: "This is what a police state is like"', The Spectator, 30 March 2020 https://www.spectator.co.uk/article/former-supreme-court-justice-this-is-what-a-police-state-is-like-

'Local authority in Wales uses drones to tell public to stay at home', Urban Air Mobility, 30 March 2020 https://www.urbanairmobilitynews.com/first-responders/drone-epidemic-reaches-wales-during-the-coronavirus-pandemic/

Woollacott, Emma, 'Tracking the trackers: coronavirus surveillance around the world', Forbes, 25 March 2020 https://www.forbes.com/sites/emmawoollacott/2020/03/25/tracking-the-trackers-coronavirus-surveillance-around-the-world/#c455f9502d8f

'Questioning conventional wisdom in the Covid-19 crisis, with Dr Jay Bhattacharya', 31 March 2020, YouTube https://www.youtube.com/watch?time_continue=418&v=-UO3Wd5urg0&feature=emb_logo

Pramod, Naga, 'Digital ID, Bill Gates vaccine record, and payments system combo to be trialed in Africa', Reclaim the Net,

17 July 2020 https://reclaimthenet.org/gates-digital-id-vaccine-record-and-payments-system-tested-in-africa/amp/

'"Sobriety tags" come into force', Ministry of Justice, GOV.UK, 19 May 2020 https://www.gov.uk/government/news/sobriety-tags-come-into-force

Stern, Christopher, Oster, Shai and Gold, Ashley, 'Coronavirus spurs demand for touchless technology', The Information, 26 March 2020 https://www.theinformation.com/articles/coronavirus-spurs-demand-for-touchless-technology

Burt, Chris, '"Smile to Pay" facial recognition system now at 300 locations in China', Biometric Update, 16 November 2018 https://www.biometricupdate.com/201811/smile-to-pay-facial-recognition-system-now-at-300-locations-in-china

Vega, Nicolas, 'Amazon tests Whole Foods payment system that uses hands as ID', New York Post, 3 September 2019 https://nypost.com/2019/09/03/amazon-testing-payment-system-that-uses-hands-as-id/

Pascu, Luana, 'Apple patent leverages face vein matching for improved Face ID biometric authentication', Biometric Update, 21 July 2010 https://www.biometricupdate.com/202007/apple-patent-leverages-face-vein-matching-for-improved-face-id-biometric-authentication

FindFace https://findface.pro/en/about/

Strong, Jennifer and the MIT Technology Review, 'Who owns your face?', In Machines We Trust podcast, 12 August 2020 https://podcasts.apple.com/us/podcast/who-owns-your-face/id1523584878

Face ++ https://www.faceplusplus.com

Christian, Alex, 'Bosses started spying on remote workers. Now they're fighting back', *Wired*, 10 August 2020 https://www.wired.co.uk/article/work-from-home-surveillance-software

Hernandez, Karina, 'Even if you're working from home, your employer is still keeping track of your productivity – here's what you need to know', CNBC Make It, 20 March 2020 https://www.cnbc.com/2020/03/19/when-working-from-home-employers-are-watching---heres-what-to-know.html

Dick, Philip K., *The Minority Report and Other Classic Stories* (New York: Citadel Press, 2002)

'Advances in AI are used to spot signs of sexuality', *Economist*, 9 September 2017 https://www.economist.com/science-and-technology/2017/09/09/advances-in-ai-are-used-to-spot-signs-of-sexuality

Lewis, Paul, '"I was shocked it was so easy": meet the professor who says facial recognition can tell if you're gay', *Guardian*, 7 July 2018 https://www.theguardian.com/technology/2018/jul/07/artificial-intelligence-can-tell-your-sexuality-politics-surveillance-paul-lewis

'Conversation with Yuval Noah Harari', SayIt, National Government of Taiwan, 30 June 2020 https://sayit.pdis.nat.gov.tw/2020-06-30-conversation-with-yuval-noah-harari

Biello, David and Tang, Audrey, 'How digital innovation can fight pandemics and strengthen democracy', TED (June 2020) https://www.ted.com/talks/audrey_tang_how_digital_innovation_can_fight_pandemics_and_strengthen_democracy#t-5882

Tiliakou, Katerina, 'From competition to competency: Taiwan looks to revamp its education system', TRTWorld, 24 December 2019 https://www.trtworld.com/perspectives/from-competition-to-competency-taiwan-looks-to-revamp-its-education-system-32438

Chapter 3: Creating You

Hornstein, Gail A., *Agnes's Jacket: A Psychologist's Search for the Meanings of Madness* (Emmaus, PA: Rodale Books, 2012)

Hunter, Clare, *Threads of Life: A History of the World Through the Eye of a Needle* (London: Sceptre Books, 2019)

Busetta, Laura, 'The self-portrait in digital media: repetition, manipulation, update', in Tinel-Temple, Muriel, Busetta, Laura and Monteiro, Marlène (eds), *From Self-Portrait to Selfie: Representing the Self in the Moving Image* (Oxford: Peter Lang, 2019), pp. 179–202

Fang, Taylor, 'What do adults not know about my generation and technology?', MIT *Technology Review, The Youth Issue*, vol. 123 (January/February 2020)

'Marshall McLuhan 1977 interview – violence as a quest for identity', 5 June 2016, YouTube https://www.youtube.com/watch?v=ULI3x8WIxus

Coulmas, Florian, *Identity: A Very Short Introduction* (Oxford: Oxford University Press, 2019)

Dray, Kayleigh, 'Comparison culture: It's OK to be irrationally jealous as you scroll through Instagram', *Stylist* (April 2020) https://www.stylist.co.uk/life/instagram-jealousy-comparison-culture-envy-how-to-curb-feelings/375756

Petrarca, Emilia, 'Body con job', *The Cut*, 14 May 2018 https://www.thecut.com/2018/05/lil-miquela-digital-avatar-instagram-influencer.html

Crous, Marisa, 'Meet the world's first digital supermodel', W24, 16 August 2018 https://www.news24.com/w24/style/fashion/trends/meet-the-worlds-first-digital-supermodel-20180815

Jadezweni, Afika, 'Is SA's first computer generated influencer our first peek into life after Covid-19?' W24, 24 April 2020 https://www.news24.com/w24/style/fashion/is-sas-first-cgi-influencer-our-first-peek-into-how-industries-may-be-digitalised-after-covid-19-20200424

The Diigitals Model Agency https://www.thediigitals.com/models

Jackson, Lauren Michele, 'Shudu Gram is a white man's digital projection of real-life black womanhood', *New Yorker*, 4 May 2018 https://www.newyorker.com/culture/culture-desk/shudu-gram-is-a-white-mans-digital-projection-of-real-life-black-womanhood

Toschi, Deborah and Villa, Federica, 'From mass media studies to self-produced media studies: strategies of self-portraiture in pregnancy and video diaries', in Tinel-Temple, Busetta and Monteiro, Marlène (eds), *From Self-Portrait to Selfie*, pp. 149–78

The Children's Commissioner for England https://www.childrenscommissioner.gov.uk/digital/who-knows-what-about-me/

Dodson, P. Claire, '"The Sims" Is the perfect game for social distancing during the coronavirus pandemic', *Teen Vogue*, 19 March

2019 https://www.teenvogue.com/story/the-sims-perfect-game
-social-distancing-coronavirus

Minotti, Mike, 'The Sims 4 has reached 20 million players',
VentureBeat, 30 January 2020 https://venturebeat.com/2020/
01/30/the-sims-4-has-reached-20-million-players/

Owens, Cassie, '"The fantasy is just ordinary life": How gamers
are returning to "The Sims" to escape the pandemic', Philadelphia
Inquirer, 15 April 2020 https://www.inquirer.com/life/
coronavirus-covid-pandemic-fantasy-the-sims-animal-crossing-
games-jane-mcgonigal-20200415.html

Jamison, Leslie, 'The digital ruins of a forgotten future', Atlantic,
(December 2017) https://www.theatlantic.com/magazine/
archive/2017/12/second-life-leslie-jamison/544149/

Roble, Doug, 'Digital humans that look just like us', TED
(April 2019) https://www.ted.com/talks/doug_roble_
digital_humans_that_look_just_like_us

Yee, Nick and Bailenson, Jeremy, 'The Proteus Effect: The
effect of transformed self-representation on behavior',
Human Communication Research, vol. 33, no. 3 (1 July 2007),
pp. 271–90

Chapter 4: Connecting You
Harari, Yuval Noah, 'Yuval Noah Harari on big data, Google and
the end of free will', Financial Times, 26 August 2016 https://www.
ft.com/content/50bb4830-6a4c-11e6-ae5b-a7cc5dd5a28c

'Keynote (Google I/O '18)', 8 May 2018, YouTube https://www.
youtube.com/watch?v=ogfYd705cRs

Hern, Alex, 'Google's "deceitful" AI assistant to identity itself as a robot during calls', *Guardian*, 11 May 2018, https://www.theguardian.com/technology/2018/may/11/google-duplex-ai-identify-itself-as-robot-during-calls

el Kaliouby, Rana, *Girl Decoded: My Quest to Make Technology Emotionally Intelligent — and Change the Way We Interact Forever* (London: Penguin Business, 2020)

Affectiva https://www.affectiva.com

Tweed, Adam, 'From Eliza to Ellie: the evolution of the AI therapist', AbilityNet, 14 May 2019 https://abilitynet.org.uk/news-blogs/eliza-ellie-evolution-ai-therapist

'Friendship', YouGov survey, July 2019 https://d25d2506sfb94s.cloudfront.net/cumulus_uploads/document/m97e4vdjnu/Results%20for%20YouGov%20RealTime%20(Friendship)%20164%205.7.2019.xlsx%20%20%5BGroup%5D.pdf

Michel, Arthur Holland, 'Interview: The professor of robot love', The Centre for the Study of the Drone at Bard College, 25 October 2013 https://dronecenter.bard.edu/interview-professor-robot-love/

Armstrong, Doree, 'Emotional attachment to robots could affect the outcome on battlefield', University of Washington News, 17 September 2013 https://www.washington.edu/news/2013/09/17/emotional-attachment-to-robots-could-affect-outcome-on-battlefield/

Lin, Patrick, 'Relationships with robots: good or bad for humans?', *Forbes*, 1 February 2016 https://www.forbes.com/

sites/patricklin/2016/02/01/relationships-with-robots-good
-or-bad-for-humans/#d6cf0567adc2

'Eugenia Kuyda 1', 15 August 2019, YouTube https://www.
youtube.com/watch?v=BptEOam30bk&list=PLUKWn7ur3662
g6x8D9F-Wi0as64yLXRuX&index=9&t=0s

'John Macinnes on virtual beings and David Bowie',
15 August 2019, YouTube https://www.youtube.com/watch?
v=4-Mbq3mF-0k&list=PLUKWn7ur3662g6x8D9F-Wi0as64y
LXRuX&index=12&t=0

Lakritz, Talia, 'Healthcare workers are taping photos of themselves
to the protective gear to help put Covid-19 patients at ease',
Insider, 8 April 2020 https://www.insider.com/coronavirus
-doctors-photos-over-protective-gear-2020-4

'Neon artificial humans: How do you make a virtual being?'
29 February 2020, YouTube https://www.youtube.com/
watch?v=ZSt1qgmmdAk

Rosebud AI https://www.rosebud.ai

'New AI app gives you 25,000 free stock photos, with the
option to use any face on any body', Digital Synopsis, https://
digitalsynopsis.com/design/free-ai-generated-stock-photos/

'Edward Saatchi, Fable, kicks off the Virtual Beings Summit
LA', 12 February 2020, YouTube https://www.youtube.com/
watch?v=Ixt5xdClJXk&list=PLUKWn7ur3661TC3dXH2-
_1ZPlqmy9Y1jD&index=2&t=6s

Kei, 'Hey waifu, I'm home: Gatebox the kawaii 3D virtual
assistant', DeJapan, 20 August 2018 http://blog.dejapan.

com/2018/08/20/gatebox-ai-virtual-waifu-home-assistant-robot/

'Japanese man "marries" hologram character Hatsune Miku', 20 November 2018, YouTube https://www.youtube.com/watch?v=dtu4t_Zc3d4

Azuma, Hiroki, *Otaku: Japan's Database Animals*, translated by Jonathan E. Abel and Shion Kono (Minneapolis: University of Minnesota Press, 2009)

Deahl, Dani, 'Grimes posted the music stems and video files for her single so anyone can remix it', The Verge, 2 April 2020 https://www.theverge.com/2020/4/2/21205769/grimes -music-stems-video-files-wetransfer-youll-miss-me-when-im-not-around-remix-coronavirus

Harney, Matt, 'The evolution of mixed reality avatars', Hackernoon, 13 December 2018 https://medium.com/hackernoon/the-evolution-of-mixed-reality-avatars-690247d9e7d0

'Hatsune Miku', Wikipedia (13 November 2020) https://en.wikipedia.org/wiki/Hatsune_Miku

'ConVRself: Self-counselling through immersive virtual reality', Freud-me, Virtual Bodyworks http://www.freud-me.com

de Botton, Alain, *The Consolations of Philosophy* (London: Penguin, 2001)

Chapter 5: Replacing You

Locke, John, *An Essay Concerning Human Understanding*, edited by Peter H. Nidditch (New York: Oxford University Press, 1998)

Bardsley, Charles Wareing, *A Dictionary of English and Welsh Surnames: With Special American Instances* (London: Oxford University Press, 1901) http://archive.org/stream/adictionaryengl00goog# page/n6/mode/2up

Bughin, Jacques; Hazan, Eric; Lund, Susan; Dahlström, Peter; Wiesinger, Anna; and Subramaniam, Amresh, 'Skill shift: Automation and the future of the workforce', McKinsey, 23 May 2018 https://www.mckinsey.com/ featured-insights/future-of-work/skill-shift-automation-and-the -future-of-the-workforce

Desjardins, Jeff, '7 charts on the future of automation', World Economic Forum, 25 February 2019 https://www.weforum. org/agenda/2019/02/the-outlook-for-automation-and -manufacturing-jobs-in-seven-charts

Open AI https://openai.com

Asparouhov, Delian, 'Quick thoughts on GPT3', Operators & Delian's Ramblings, 18 July 2020 https://delian.substack. com/p/quick-thoughts-on-gpt3

'Balaji Srinivasan: applications: today & 2025', 20 May 2020, YouTube https://www.youtube.com/watch?time_continue= 142&v=3jPYk7ucrjo&feature=emb_logo

Pearson, Ian, *You Tomorrow* (self-published, 2013)

Yarow, Jay, 'Sergey Brin: "We want Google to be the third half of your brain"', Business Insider, 8 September 2010 https://www. businessinsider.com/sergey-brin-we-want-google-to-be-the- third-half-of-your-brain-2010-9?r=US&IR=T

'SFBW19 – The pseudonymous economy – Balaji Srinivasan', 14 November 2019, YouTube https://www.youtube.com/watch?v=Dur918GqDIw

Jaynes, Tyler L., 'Legal personhood for artificial intelligence: citizenship as the exception to the rule', *AI & Society*, vol. 35 (June 2020), pp. 343–354

'ARISE – entrevista – David J. Gunkel', 10 July 2020, YouTube https://www.youtube.com/watch?v=8PAxjvPV9lo&feature=youtu.be

Gunkel, David J., *The Machine Question: Critical Perspectives on AI, Robots, and Ethics* (Cambridge, MA: MIT Press, 2012)

Gunkel, David J., *Robot Rights* (Cambridge, MA: MIT Press, 2018)

Prescott, Tony J., 'Robots are not just tools', *Connection Science*, vol. 29, no. 2 (19 April 2017), pp. 142–49

McGee, Patrick, 'Elon Musk-backed Neuralink unveils brain-implant technology', Financial Times, 17 July 2019 https://www.ft.com/content/144ba3b4-a85a-11e9-984c-fac8325aaa04

Musk, Elon and Neuralink, 'An integrated brain-machine interface platform with thousands of channels', 16 July 2019 https://www.documentcloud.org/documents/6204648-Neuralink-White-Paper.html

Smith, Thomas and Chiang, Chris, 'A Scientist and Engineer Explain Everything Elon Musk's Neuralink Can (and Can't) Do', OneZero, Medium (September 2017) https://onezero.medium.com/a-scientist-and-engineer-explain-everything-elon-musks-neuralink-can-and-can-t-do-68e7f5325e09

Fan, Shelly, 'DARPA's new project is investing millions in brain-machine interface tech', SingularityHub, 5 June 2019 https://singularityhub.com/2019/06/05/darpas-new-project-is-investing-millions-in-brain-machine-interface-tech/

Griffin, Matthew, 'DARPA human telepathy project will let people transmit images into other people's brains, Fanatical Futurist, 19 September 2019 https://www.fanaticalfuturist.com/2019/09/new-darpa-neuro-tech-project-will-let-people-telepathically-transmit-images-to-other-peoples-brains/

Eagleman, David, The Brain: The Story of You (London: Canongate Books, 2015)

Eagleman, David, Livewired: The Inside Story of the Ever-Changing Brain (London: Canongate Books, 2020)

'Rewiring the brain w/David Eagleman | FUTURES Podcast LIVE', 2 September 2020, YouTube https://www.youtube.com/watch?v=kD-yvR7W50k

Watts, Peter, 'Hive consciousness', Aeon, 27 May 2015 https://aeon.co/essays/do-we-really-want-to-fuse-our-brains-together

Transhumanist Party http://www.zoltanistvan.com/Transhumanist Party.html

US Transhumanist Party https://transhumanist-party.org

Chapter 6. Enhancing You

Perry, John (ed.), Personal Identity (Berkeley: University of California Press, 2008)

Williams, Bernard, 'The self and the future', *The Philosophical Review*, vol. 79, no. 2. (April 1970), pp. 161–81

'Chris Dancy – CyborgCamp MIT 2014', 10 October 2014, YouTube https://www.youtube.com/watch?v=JR25HPprBGw

'Submission to the House of Commons Science and Technology Committee inquiry on commercial genomics', Nuffield Council on Bioethics, 26 April 2019 https://www.nuffieldbioethics.org/assets/pdfs/Nuffield-response-to-ST-Committee-inquiry-on-commercial-genomics-April-2019-FINAL.pdf

'Whole genome sequencing of babies', Nuffield Council on Bioethics, 27 March 2018 https://www.nuffieldbioethics.org/publications/whole-genome-sequencing-of-babies

'Genome editing: an ethical review', Nuffield Council on Bioethics, 30 September 2016 https://www.nuffieldbioethics.org/publications/genome-editing-an-ethical-review

Pippa Campbell Health https://www.pippacampbellhealth.com

'The NHS Long Term Plan', NHS (August 2019) https://www.longtermplan.nhs.uk/wp-content/uploads/2019/08/nhs-long-term-plan-version-1.2.pdf

'The 100,000 genomes project', Genomics England, https://www.genomicsengland.co.uk/about-genomics-england/the-100000-genomes-project/

'Hacking Darwin: genetic engineering and the future of humanity', 20 May 2020, YouTube https://www.youtube.com/watch?v=nQSywF4nVQQ

'Prof Joyce Harper | genetics debate | proposition', 31 May 2018, YouTube https://www.youtube.com/watch?v=Zne2gx-q_mM

Metzl, Jamie, *Hacking Darwin: Genetic Engineering and the Future of Humanity* (Naperville, IL: Sourcebooks, 2020)

Tangermann, Victor, 'A gene-editing experiment on human embryos went horribly wrong', Futurism, 17 June 2020 https://futurism.com/neoscope/gene-editing-human-embryos-went-horribly-wrong

Dick, Samantha, 'On this day: a pregnant man gives birth for the first time in history', The New Daily, 29 June 2020 https://thenewdaily.com.au/news/national/2020/06/29/on-this-day-pregnant-man-gives-birth/

Sebag-Montefiore, Clarissa, 'Is motherhood gendered?', Aeon, 7 September 2016 https://aeon.co/essays/when-trans-people-become-parents-who-gets-called-mother

Rothblatt, Martine, *From Transgender to Transhuman: A Manifesto on The Freedom of Form* (self-published, 2011)

'Prof. Steve Fuller on transhumanism: ask yourself what is human?', 25 August 2019, YouTube https://www.youtube.com/watch?v=ndDBvGFUvWw

'Dawn of the transhuman era w/ prof Steve Fuller | FUTURES podcast LIVE', 24 June 2020, YouTube https://www.youtube.com/watch?v=Cnryg3h2bxo

'The future of human augmentation 2020: opportunity or dangerous dream?', Kaspersky Lab (2020) https://media.kasperskydaily.com/wp-content/uploads/sites/86/2020/09/17130024/Kaspersky-The-Future-of-Human-Augmentation-Report.pdf

Currie, Richard, 'As we stand on the precipice of science fiction into science fact, people say: Hell yeah, I want to augment my eyesight!', The Register, 18 September 2020 https://www. theregister.com/AMP/2020/09/18/people_want_to_be_ cyborgs_kaspersky/?__twitter_impression=true

Minelle, Bethany, 'Elon Musk and Grimes change baby X Æ A-12's name due to Californian law', Sky News, 25 May 2020 https:// news.sky.com/story/elon-musk-and-grimes-change-baby-x-ae-a -12s-name-due-to-californian-law-11994445

Connor, Steve, 'Professor has world's first silicon chip implant', Independent, 26 August 1998 https://www.independent.co.uk/ news/professor-has-worlds-first-silicon-chip-implant-1174101. html

Mason, Luke Robert, 'Cyborg experiments w/ prof. Kevin Warwick', Episode 1, Futures podcast (2019) https://futures podcast.net/episodes/01-kevinwarwick

'The internet of bodies is here: tackling new challenges of technology governance', World Economic Forum, 6 August 2020 https://www.weforum.org/reports/the-internet-of-bodies-is -here-tackling-new-challenges-of-technology-governance

Zoltan Istvan http://www.zoltanistvan.com

Gohd, Chelsea, 'Can we genetically engineer humans to survive missions to Mars?' Space.com, 7 November 2019 https://www. space.com/genetically-engineer-astronauts-missions-mars- protect-radiation.html

Bittel, Jason, 'Tardigrade protein helps human DNA withstand radiation', Nature, 20 September 2016 https://www.nature.

com/news/tardigrade-protein-helps-human-dna-withstand-radiation-1.20648

'Read Yuval Harari's blistering warning to Davos in full', World Economic Forum

Wee, Sui-Lee, 'China is collecting DNA from tens of millions of men and boys, using US equipment', *New York Times*, 30 July 2020 https://www.nytimes.com/2020/06/17/world/asia/China-DNA-surveillance.html

Dirks, Emilie and Leibold, Dr James, 'Genomic surveillance', Australian Strategic Policy Institute, 17 June 2020 https://www.aspi.org.au/report/genomic-surveillance

'Universal declaration on the human genome and human rights', United Nations Human Rights Office of the High Commissioner, 11 November 1997 https://www.ohchr.org/EN/ProfessionalInterest/Pages/HumanGenomeAndHumanRights.aspx

Chapter 7: Destroying You

Scruton, Roger, *An Intelligent Person's Guide to Philosophy* (London: Duckworth, 1996)

Advantageous, dir. Jennifer Phang, Netflix, 2015

Srinivasan, Balaji S., 'The purpose of technology', Balajis.com, 19 July 2020 https://balajis.com/the-purpose-of-technology/

Fortson, Danny, 'Longevity Fund's Laura Deming: "Age is a disease"', *Danny in The Valley* podcast, 12 October 2018 https://play.acast.com/s/dannyinthevalley/lauradenning

UNITY Biotechnology website https://unitybiotechnology.com

Swaminathan, Nikhil, 'A Silicon Valley scientist and entrepreneur who invented a drug to explode double chins is now working on a cure for aging', Quartz, 6 January 2017 https://qz.com/878446/unity-biotechnology-cure-for-aging/

de Grey, Aubrey, 'A roadmap to end aging', TED (2005) https://www.ted.com/talks/aubrey_de_grey_a_roadmap_to_end_aging#t-933003

Fortson, Danny, 'SENS foundation's Aubrey de Grey: "You may live for a million years"', Danny in the Valley podcast, 16 February 2018 https://play.acast.com/s/dannyinthevalley/sensfoundationsaubreydegrey-youmayliveforamillionyears-

Kelland, Kate, 'Who wants to live forever? Scientist sees aging cured', Reuters, 4 July 2011 https://www.reuters.com/article/idUSTRE7632ID20110704?irpc=932

Winarsky, Hanne, 'The biology of aging', Bio Eats World podcast, 22 September 2020 https://podcasts.apple.com/gb/podcast/bio-eats-world/id1529318900?i=1000492084217

'FM-2030', Wikipedia https://en.wikipedia.org/wiki/FM-2030

Terasem Movement Inc. https://terasemcentral.org/about.html

Darwin, Mike, 'Does personal identity survive cryopreservation?' Chronosphere, 23 February 2011 http://chronopause.com/chronopause.com/index.php/2011/02/23/does-personal-identity-survive-cryopreservation/index.html

Danaher, John, 'Minerva on the ethics of cryonics', Episode 46, Philosophical Disquisitions podcast, 5 October 2018 https://

philosophicaldisquisitions.blogspot.com/2018/10/episode-46
-minerva-on-ethics-of-cryonics.html?utm_source=feedburner
&utm_medium=feed&utm_campaign=Feed:+philosophical
discursions+(Philosophical+Discursions)

Upload: Season 1, dir. Greg Daniels, Amazon Prime Video, 2020

Rothblatt, Martine, Virtually Human: The Promise – and the Peril – of
Digital Immortality (London: Picador, 2015)

'Bina48 + Bruce Duncan – diversity in AI', 21 June 2018,
YouTube https://www.youtube.com/watch?v=089UCS6BrGQ

'James Vlahos, HereAfter AI, on Legacy Avatars – life after
life', 28 July 2020, YouTube https://www.youtube.com/
watch?v=dxXftWfJC_8&list=PLUKWn7ur3660sx9xsCEZHc
-5QXksqye_H&index=18&t=0s

Gyu-lee, Lee, '"Meeting You" creator on his controversial show:
"I hope it opens up dialogue"', Korea Times, 5 April 2020 https://
www.koreatimes.co.kr/www/art/2020/04/688_287372.html

Goode, Lauren, 'Inside 8i, the VR company trying to take
holographic videos to the mainstream', The Verge, 13 February
2017 https://www.theverge.com/2017/2/13/14600662/8i-
holograms-ar-vr-mobile-app-holo-google-tango

'Virtual Beings Summit 2020', YouTube playlist https://www.
youtube.com/playlist?list=PLUKWn7ur3660sx9xsCEZHc
-5QXksqye_H

Stein, Scott, 'Deepak Chopra made a digital clone of himself, and
other celebs could soon follow', Cnet, 5 December 2019 https://
www.cnet.com/google-amp/news/deepak-chopra-made-
a-digital-clone-of-himself-and-other-celebs-could-soon-follow/

'The Times: JFK Unsilenced by ROTHCO', The Drum, 2018
https://www.thedrum.com/creative-works/project/rothco
-the-times-jfk-unsilenced

Fuller, Steve, Nietzschean Meditations: Untimely Thoughts at the Dawn of
the Transhuman Era (Basel: Schwabe Verlagsgruppe, 2019)

Walbesser, Jordan L., 'Finding meaning in the death of virtual
identities', Buffalo Intellectual Property Law Journal, vol. 10 (2014)
https://digitalcommons.law.buffalo.edu/buffaloipjournal/
vol10/iss1/3

BIBLIOGRAPHY

Adams, Scott, *Win Bigly: Persuasion in a World Where Facts Don't Matter* (New York: Penguin, 2017)

Azuma, Hiroki, *Otaku: Japan's Database Animals*, translated by Jonathan E. Abel and Shion Kono (Minneapolis: University of Minnesota Press, 2009)

Bailenson, Jeremy, *Experience on Demand: What Virtual Reality Is, How it Works, and What it Can Do* (New York: W. W. Norton & Company, 2019)

Bailey, David Evans, *Virternity: The Quest for a Virtual Eternity: A Treatise on the Aims and Goals of the Virternity Project* (self-published, 2017)

Bardsley, Charles Wareing, *A Dictionary of English and Welsh Surnames: With Special American Instances* (London: Oxford University Press, 1901)

Birch, David, *Identity Is the New Money* (London: London Publishing Partnership, 2014)

Blakemore, Sarah-Jayne, *Inventing Ourselves: The Secret Life of the Teenage Brain* (New York: Doubleday, 2018)

Boden, Margaret A., *AI: Its Nature and Future* (Oxford: Oxford University Press, 2016)

Cameron, Kim, 'The Laws of Identity', www.identityblog.com (2005)

Casey, Michael J. and Vigna, Paul, *The Truth Machine* (London: HarperCollins, 2018)

Coulmas, Florian, *Identity: A Very Short Introduction* (Oxford: Oxford University Press, 2019)

Denison, Rayna, *Anime: A Critical Introduction* (London: Bloomsbury Academic, 2015)

Devlin, Kate, *Turned On: Science, Sex and Robots* (London: Bloomsbury Sigma, 2018)

Diamandis, Peter H. and Kotler, Steven, *The Future Is Faster Than You Think: How Converging Technologies Are Transforming Business, Industries, and Our Lives* (London: Simon & Schuster, 2020)

Dick, Philip K., *The Minority Report and Other Classic Stories* (New York: Citadel Press, 2002)

Eagleman, David, *The Brain: The Story of You* (London: Canongate Books, 2015)

Eagleman, David, *Livewired: The Inside Story of the Ever-Changing Brain* (London: Canongate Books, 2020) el Kaliouby, Rana, *Girl Decoded: My Quest to Make Technology Emotionally Intelligent – and Change the Way We Interact Forever* (London: Penguin Business, 2020)

Fuller, Steve, *Nietzschean Meditations: Untimely Thoughts at the Dawn of the Transhuman Era* (Basel: Schwabe Verlagsgruppe, 2019)

Goffman, Erving, *The Presentation of Self in Everyday Life* (New York: Doubleday, 1959)

Gunkel, David J., *The Machine Question: Critical Perspectives on AI, Robots, and Ethics* (Cambridge, MA: MIT Press, 2012)

Gunkel, David J., *Robot Rights* (Cambridge, MA: MIT Press, 2018)

Harari, Yuval Noah, *Sapiens: A Brief History of Humankind* (London: Penguin, 2011)

Hornstein, Gail A., *Agnes's Jacket: A Psychologist's Search for the Meanings of Madness* (Emmaus, PA: Rodale Books, 2012)

Hunter, Clare, *Threads of Life: A History of the World Through the Eye of a Needle* (London: Sceptre Books, 2019)

Kind, Amy, *Persons and Personal Identity* (Cambridge: Polity Press, 2015)

Kurzweil, Ray, *The Singularity Is Near: When Humans Transcend Biology* (London: Duckworth, 2006)

Locke, John, *An Essay Concerning Human Understanding*, edited by Peter H. Nidditch (New York: Oxford University Press, 1998)

Metzl, Jamie, *Hacking Darwin: Genetic Engineering and the Future of Humanity* (Naperville, IL: Sourcebooks, 2020)

Murray, Douglas, *The Madness of Crowds: Gender, Race and Identity* (London: Bloomsbury, 2019)

Perry, John (ed.), *Personal Identity* (Berkeley: University of California Press, 2008)

Plato, *Complete Works*, edited by J. M. Cooper (Indianapolis: Hackett Publishing, 1997)

Preukschat, Alex and Reed, Drummond, *Self-Sovereign Identity: Decentralized Digital Identity and Verifiable Credentials*, Version 2 (New York: Manning Publications, 2020)

Rorty, Amélie Oksenberg (ed.), *The Identities of Persons* (Berkeley: University of California Press, 1976)

Rothblatt, Martine, *From Transgender to Transhuman: A Manifesto on the Freedom of Form* (self-published, 2011)

Rothblatt, Martine, *Virtually Human: The Promise — and the Peril — of Digital Immortality* (London: Picador, 2015)

Scruton, Roger, *An Intelligent Person's Guide to Philosophy* (London: Duckworth, 1996)

Sontag, Susan, *At the Same Time* (London: Penguin, 2008)

Busetta, Laura, 'The self-portrait in digital media: repetition, manipulation, update', in Tinel-Temple, Muriel, Busetta, Laura and Monteiro, Marlène (eds), *From Self-Portrait to Selfie: Representing the Self in the Moving Image* (Oxford: Peter Lang, 2019)

INDEX